Gene Doping – The Future of Doping?

Swen Körner/Stefanie Schardien/
Birte Steven-Vitense/Steffen Albach/
Edgar Dorn/Tobias Arenz/Marcel Scharf

Gene Doping – The Future of Doping?

Teaching Unit
Gene Doping in Competitive Sports

Bibliographic Information published by the Deutsche Nationalbibliothek
The Deutsche Nationalbibliothek lists this publication in the Deutsche
Nationalbibliografie; detailed bibliographic data is available in the internet
at http://dnb.d-nb.de.

Library of Congress Cataloging-in-Publication Data
Names: Körner, Swen, author.
Title: Gene doping - the future of doping? : teaching unit, gene doping in competitive sports / Swen Körner [and six others].
Description: Frankfurt am Main : Peter Lang Gmbh, 2016. | Includes bibliographical references.
Identifiers: LCCN 2016007904| ISBN 9783631670941 (print) | ISBN 9783653063462 (ebook)
Subjects: LCSH: Doping in sports—Moral and ethical aspects. | Doping in sports—Law and legislation. | Sports medicine—Moral and ethical aspects.
Classification: LCC RC1230 .K67 2016 | DDC 362.29/088796—dc23 LC record available at http://lccn.loc.gov/2016007904

SPONSORED BY THE

 Federal Ministry
of Education
and Research

Sponsored by: Federal Ministry of Education and Research
(Bundesministerium für Bildung und Forschung -BMBF)

Text and editing: (Project team): Prof. Dr. Swen Körner, German Sport University Cologne; Dr. Stefanie Schardien; Birte Steven, German Sport University Cologne; Edgar Dorn, Hildesheim University Foundation; Tobias Arenz, German Sport University Cologne; Marcel Scharf, German Sport University Cologne; Steffen Albach, buisness/ sports teacher.

Collaborators: Dr. Uta Bittner, University Ulm; Prof. Dr. Franz Bockrath, TU Darmstadt; Sarah Breitbach, University of Mainz; Hendrik Forster, University of Mainz; Ilke Glockentöger, University of Oldenburg; Dr. Thorsten Heinemann, University of Frankfurt; Katharina Lammert, National Anti Doping Agency Germany - NADA Deutschland; Dr. Dorothea Magnus, University of Hamburg; Stefanie Mosler, German Sport University Cologne; Elmo Neuberger, University of Mainz; Dr. Annika Steinmann, German Sport University Cologne; Dr. Martina Velders, University of Ulm.

ISBN 978-3-631-67094-1 (Print)
E-ISBN 978-3-653-06346-2 (E-Book)
DOI 10.3726/978-3-653-06346-2

© Peter Lang GmbH
Internationaler Verlag der Wissenschaften
Frankfurt am Main 2016
All rights reserved.
Peter Lang Edition is an Imprint of Peter Lang GmbH.

Peter Lang – Frankfurt am Main · Bern · Bruxelles · New York ·
Oxford · Warszawa · Wien

All parts of this publication are protected by copyright. Any
utilisation outside the strict limits of the copyright law, without
the permission of the publisher, is forbidden and liable to
prosecution. This applies in particular to reproductions,
translations, microfilming, and storage and processing in
electronic retrieval systems.

This publication has been peer reviewed.
www.peterlang.com

Table of Contents

Preface 9

User Notes 11

Teacher materials: Volume 1

Chapter 1 Overview 15

Chapter 2 Topic relevance 17
    Gene doping in competitive sport 17

Chapter 3 Background information for the teacher
               (Subject matter analysis) 19
    3.1 Gene doping definition 19
    3.2 Medicine 20
    3.3 Ethics 21
    3.4 Law 22

Chapter 4 Curricular tie-ins 25

Chapter 5 Options for using the teaching unit 29
    5.1 Modular options 29
        5.1.1 Focus: Scientific foundations of gene doping 29
        5.1.2 Focus: Legal foundations of gene doping 29
        5.1.3 Focus: Ethical aspects of gene doping 30
    5.2 Alternative option 31
    5.3 Classical option 32
        5.3.1 Teaching unit schematic overview 32

    5.3.2 Course planning 33
    5.3.3 Detailed guidelines for the individual phases................................. 35
          Phase 1: Introduce and raise awareness of the issue 35
          Phase 2: Scientific part 36
          Phase 3 (optional): In-depth scientific part 36
          Phase 4: Scenario (news article) and preparing for the
                       crisis talk... 37
          Phase 5: Crisis talk... 38
          Phase 6: Appraising the crisis talk... 44
          Phase 7: Conclusion 45

Chapter 6 Additional readings 47

Chapter 7 Annex 49

7.1 Material for Phase 1 49

7.2 Material for Phase 2 54
    Scientific basis of gene doping 54
    Evaluation rubric 57

7.3 Material for Phase 3 (optional): In-depth scientific section 58
    In-depth text 1: Indirect detection of gene doping................................. 58
    Evaluation rubric 59
    Evaluation rubric 61

7.4 Material for Phase 4 67
    Lead-in to the crisis talk using the scenario (handout) 67
    Evaluation rubric: Lawyer.. 67
    Evaluation rubric: Spectator/Fan 71
    Evaluation rubric: Athlete/Marco Epowitz... 72
    Evaluation rubric: coaches... 73
    Evaluation rubric: federation official.. 74

Chapter 8 Additional material slide presentation:
legal basics of gene doping 75
          Compact slide overview 75

Student materials: Volume 2

Chapter 1 Scientific basics of gene doping 87
MATERIAL 1: Basic text 87
MATERIAL 2: in-depth text 1 90
MATERIAL 3: In-depth text 2 91

Chapter 2 Crisis talk material 95
MATERIAL 4: News report for the crisis talk 95

Chapter 3 Infopack for the lawyer role 97
MATERIAL 5: Introduction to the lawyer's role 97
MATERIAL 6: Gene doping and the law 100
MATERIAL 7: Legal dimensions of gene doping 102

Chapter 4 Infopack for the fan/spectator role 107
MATERIAL 8: Introduction to the fan role 107
MATERIAL 9: The F.A.Z. in conversation with Prof. K.-H. Bette ... 108
MATERIAL 10: The following are excerpts from talks with two well-known experts on the subject of "gene doping in sport" 109
MATERIAL 11: "Unrestricted doping is fairer" 110

Chapter 5 Infopack for the athlete role 113
MATERIAL 12: Introduction to the athlete role 113
MATERIAL 13: Athlete 113
MATERIAL 14: Discussion structuring help 120

Chapter 6 Infopack for the role of coach 121

   MATERIAL 15: Introduction to the role of coach 121

   MATERIAL 16: Excerpts from interviews with two well-known
                      experts on "gene doping in sport" 122

   MATERIAL 17: Andy Miah on ethical reflections on gene doping 123

   MATERIAL 18: "Unrestricted doping is fairer" 123

   MATERIAL 19: "Gold in the genes" 126

Chapter 7 Infopack for the role of federation official 127

   MATERIAL 20: Introduction to the role of federation official 127

   MATERIAL 21: The German Olympic Sports Confederation (DOSB)
                      and its self-image 127

   MATERIAL 22: The state's goals for competitive sports 129

   MATERIAL 23: Excerpts from the DOSB "State Goals for Sport
                      position paper" 129

   MATERIAL 24: Information about athlete whereabouts
                      and accessibility 130

   MATERIAL 25: Op-ed on doping in the F.A.Z. 131
                      The competitive athlete made of glass 131

Chapter 8 Supplemental material 133

   MATERIAL 26: Audience assignments 133

   MATERIAL 27: Your bottom line 134

   MATERIAL 28: Quiz 134

   MATERIAL 29: Glossary 139

Table of figures 151

Preface

Gene doping is a promise. For one thing, it cannot deliver on what it purports to do: according to the experts, experimenting with or using it on the athlete's body to deliberately trigger and control moleculogenetic regulatory mechanisms is a bridge too far as things stand now. To date, there has not been a single documented instance of gene doping, and the development of relevant detection methods is still in its infancy. But we also need to draw a distinction between this *not yet* and a fundamental *not*. While the former is merely a problem of technological development, hence only a question of time, the fundamental *not* is about the moral-ethical reasons why we *may not* or *should not* do what is technically feasible. We have learned from the rapid progress in the field of biomedicine over the last seven decades that questions of social acceptance are decided in the light of historically co-evolving norms. How a fundamental *not* can turn into a fundamental *shall* can be seen on the example of how we progressed from rescinding prohibitions in the area of generative reproduction to legalizing induced abortions in 1976 to qualified approval of preimplantation genetic diagnosis in 2011. The power of embedded norms to stymie technological leaps, however, also can hardly be overestimated if for no other reason than modern societies are at pains to control side effects from their medico-scientific advances. Time and again, technologies from medicine, pharmacology, or the military spill over into other areas of society where they are put to completely new uses – we need only to recall Teflon pans and jet propulsion. Modern competitive sport and its susceptibility to doping technologies is no exception in this regard. Amphetamines can be used to straighten out ADHD children and keep soldiers alert, but they can just as well spur performance-inhibiting resources in the athlete body into action. Modern elite sport also finds use for anabolic steroids or erythropoietin (EPO) – both of them developed exclusively for medico-therapeutic purposes.

The other reason that gene doping is a chimera of sorts is because it has an aura of fascination about it *even though* it cannot live up to what it promises. Science and all the experts in the world can blame the fictionalizing of hard facts and marshal arguments for its present non-feasibility – modern elite sport, it seems, could not care less. It buys readily into the vision of a push button technology that seems set to unlock the ultimate secrets of human performance enhancement in the micro environment of the genes and to spin them into gold. A few years ago, when Repoxygen™ came along as a genetic engineering method for the intramuscular application of the EPO gene, a German former track and field coach lost no time signaling an interest in using it. The cloned athlete as the real utopia of tomorrow's elite sport may be nothing more than fantasy run wild, but the mere fiction suffices for the world of competitive sport. In premature haste, in 2003 already *the World Anti-Doping Agency* moved to add gene doping to its prohibited list – without giving any specifics at first. The ingrained academic caution against indulging in speculation holds no water in sports. Ordering scientific knowledge is one thing; the logic of elite sports practice once again is something totally different.

Like the *debate on doping*, the *gene doping debate* is one of many voices. And like doping, gene doping – unquestionably charged by the magic word "gene" – is also something modern society is hooked on. Clearly, biochemists with an eye on gene doping are not going to stop analyzing bodily fluids and tinkering with new detection methods. Education and prevention, too, by no means are deterred from their well-intended work with high school athletes. Doping is also grist for the mills of those performing ethical, legal, and mass media functions: It can be reported on with gripping images and words, soberly adjudicated or discussed in terms of what it all means for sports. The fan base is not idle on the sidelines, either. It consumes sports but at the same time is appalled when it learns of doping in a sport. In doping, too, society reaps what it sows. Once started, society is forced to deal with the turmoil that it itself has conjured up.

It is these trends and problems that the present work ultimately addresses. You will find here specially adapted materials for examining the gene doping topic from scientific, ethical, and legal angles. Using it will foster a multiperspectival awareness of the problem that can serve as the basis for forming the requisite nuanced opinions.

Our special thanks to the Federal Ministry for Education and Research (BMBF) for supporting this project.

User Notes

The teaching unit consists of two parts, the *teacher's materials* and the *student materials*.

The teacher's materials have *detailed explanations, horizons of expectations*, or *pointers to potential applications* for all documents in the student binder.

The following are potential applications:

1. Teachers who are interested in offering a comprehensive overview that takes into account scientific, ethical, and legal aspects of the topic should take up the *classic option* (5–7 hours) or the *alternate option* (2–3 hours). The classic option covers all material in the student binder and may be printed out completely starting from page 91.
2. Teachers who wish to focus on their specialty, i.e., do not chiefly wish to pursue interdisciplinary objectives, can choose from among the *modular options* (2 hours and up). This will also work for teachers who, due to preparation time constraints, opt to treat the topic of gene doping in the context of their subject.
3. For an accelerated approach, all materials in the student binder can be surveyed initially to choose individual materials for customized use in the classroom.

Please feel free to contact Faul@dshs-koeln.de should questions arise about deploying the teaching unit.

**Teacher materials
Volume 1**

Chapter 1
Overview

	Key points at a glance
Class level	**Upper secondary grade**
Subject/subjects and curricular tie-ins	Indicated generally for Lower Saxony core curricula in: biology, values and norms (ethics), and sport, as well as, for example, for legal study, as seminar course, seminar topic or science-preparatory seminar.
Time requirement	• Classic option: five to seven class hours • Compact option: from two class hours • Alternative option: two to three class hours
Class size (classic option)	Minimum: 15 student Optimum: 20 students Maximum: 25 students
Prior knowledge/ experience	Students' connection to the subject of competitive sports is recommended
Principal methods	• Modified barometer teaching method • Work with text, caricatures, charts • Crisis talk (panel discussion)
Learning arrangements	Individual and/or pair work; group work, group discussion, frontal teaching
Differentiated activities	Feasible, e.g., through different roles and materials in the crisis talk and in-depth texts in the scientific part or through "alternative versions."
Technology	Projector, laptop (at least during the first class hour)
Subject-combining, interdisciplinary lessons	Yes (see above)
Field-tested material	Yes
Teaching unit objective	The teaching unit aims to engage students in a socially relevant discussion process. The attempt will be made to facilitate grounded reflection on bioethical, social, and legal issues concerning the use of gene technologies in (competitive) sport – specifically, practices of so-called gene doping – and thereby equip them to form in-depth, critical judgments.
Primary skill	Forming judgments

Chapter 2
Topic relevance

Gene doping in competitive sport

In recent years, ideas and practices aimed at achieving a healthier life style have focused attention in parts of our society on the notion of "enhancement." In competitive sport, with its inherent logic of improvement, enhancement in this sense of the term has long been a familiar phenomenon. The use of performance-enhancing substances and techniques, both actual and assumed, has been widespread for a long time, is prohibited in defined instances and is subjected to sanctions as doping. So-called gene doping, a special case of genetically engineered enhancements, is regarded as the form of performance enhancement in competitive sports with the greatest future potential and also as raising issues for our society that transcend competitive sport.

In competitive sport, a segment of modern society has differentiated itself by elevating improvement under a rigid competitive and record-seeking logic to its highest internal norm ("higher-faster-farther"). Gene doping holds the potential for continuing this improvement logic with innovative means. Even if its exact dimensions could only be predicted tentatively until now, gene doping portends the availability of a kind of "push button technology" that will enable sports to tackle the last open questions of human performance enhancement in the micromilieu of genes. Even more discreet than conventional doping, gene doping picks up where the legitimate effects of training and science (medicine, psychology, and training science) leave off as techniques for breaking through limits on athletic performance. And even more than conventional doping, gene doping raises issues of relevance to society as a whole that transcend the internal capacity of organized sports to process and take responsibility for. The pharmacological modulation of endogenous gene activity that biomedicine currently already classifies as practicable poses new problems and challenges for the institutional and governmental right of legal regulation because detection procedures are either nonexistent or costly. The possibility of genetically engineered enhancements has not only kickstarted the stalemated discussion about their prohibition, control and rationale in sports. It is also paralleled by a comprehensive social and bioethical audit of the non-material and material yardsticks governing performance and behavior that are applied to modern competitive sport and its actors. And even as it throws into sharp relief our fundamental individual and societal attitudes of how to deal with biomedical enhancement techniques, gene doping also continues to raise the question of what images, explicit or implicit, ought to govern our understanding of what it means to be human.

While medicine and molecular biology come to different conclusions about the current and future promise as well as the risks of genetically engineered enhancements in competitive sport, which are necessarily linked to progress in gene therapy research, it is especially the public media space that regularly hypes the subject into what amounts to a crisis discourse that, contrary to all academically grounded objections and qualifications, conjures up the cloned athlete as tomorrow's veritable competitive sports utopia. From this constant crisis and risk discourse we can expect not only the stylization and truncation that biomedical findings are subjected to in the media's editorial process but also repercussions on science itself. Doping research in the social sciences as well as the general ethical, legal, and social debate over enhancements for a healthy life must be informed by rigorously detailed knowledge of the issues involved, and it must be applied systematically to the discussion of gene doping that lies in the offing.

Chapter 3
Background information for the teacher (Subject matter analysis)

In what follows, in addition to explanations for the definition of gene doping, the topical content will be aligned with three scientific disciplines that are also central to the teaching unit.

3.1 Gene doping definition

The concept of "gene doping" can be used in a narrow and in a broad sense. Narrowly defined, gene doping is the transfer of genetic material (DNA or RNA) into a cell, an organ, or an organism. The delivery of the DNA or RNA takes place through the misuse of gene and cell therapeutic processes. In a broader sense, gene doping is also understood as a deliberate modulation of gene activity by other methods (e.g., the taking/administering of pharmacologic substances).

The World Anti-Doping Agency (WADA) until 2011 exclusively used the expanded understanding of the term, defining gene doping as the "non-therapeutic use of cells, genes, genetic elements, or the modulation of gene expression having the capacity to improve athletic performance." In the revised version of the prohibited list in force since 01.01.2015, the following methods with the potential to enhance sport performance are prohibited:

1. The transfer of polymers of nucleic acids or nucleic acid analogs;
2. The use of normal or genetically modified cells.

The modulation of gene expression by means of pharmacologic substances is deleted from the revised version. This development can be understood as a reaction to the definition's oft-criticized terminological fuzziness. For instance, pharmaceutical substances like anabolic steroids have also been known for a long time to enhance athletic performance by modulating gene expression. The problem of a selective definition, however, persists in the current version of the WADA prohibited list; for example, blood doping, strictly speaking, would fall under the gene doping definition since it also involves the use of cells.

A solution to the problems of defining a "gene doping" concept is not in sight. Any future discussion to this effect will also have to take into account epigenetic modifications of gene expression, an area where research is still in its infancy.

3.2 Medicine

From the perspective of biomedicine, the gene doping challenge stems above all from the fact that the rapid advance of knowledge in gene therapy and gene manipulation has led to an exponential increase in the ways in which athletes' performance capacity may be manipulated. To stay on top of these developments, the World Anti-Doping Agency (WADA) in 2003 added gene doping as a prohibited method to its anti-doping code. However, this also created the need for developing new detection methods for the doping control system.

While the priority in developing methods and substances for deliberate modulation of gene activity serves the identification of new therapeutic strategies for treating diseases, it is obvious that findings from biomedical and pharmacological research may be used or misused for purposes of gene doping. Because they limit human performance capacity, three physiological areas can be identified as the biologically relevant potential starting points for gene doping: skeletal muscle, oxygen supply, and energy supply. Within these areas exist myriad possibilities for influencing the related underlying molecular regulatory mechanisms by pharmacological and moleculobiological means. The expression of the myostatin growth factor, for example, can be manipulated on the levels of transcription, translation, post-transcriptional modification or intracellular signal transduction.

However, since methods and substances for modifying gene activity derive only from animal experiments and clinical studies, their use for purposes of performance enhancement in healthy humans harbors difficult to calculate health risks. Known side effects, e.g., immune reactions or uncontrolled cell growth, point to potentially serious damage to health that ultimately can even result in death.

Checking for compliance with gene doping prohibitions requires specific test procedures for detecting gene doping.

However, development of such detection methods, differentiated between screening systems and direct detection systems, is still in its infancy. Screening systems are used to analyze biomarkers (e.g., hematocrit) as indicators of deviations from an organism's normal physiological state. This allows detecting manipulations without having to know which substance and/or method they were achieved with. By contrast, direct detection procedures, for example, seek to identify vectors or inserted foreign genetic material. The question of whether the future development of direct methods is even feasible, given the complexity of potential points of attack for manipulation, is an ongoing controversy.

3.3 Ethics

Ethical questions and demands arise where scientific, technical, economic or cultural developments confront the individual or society with new challenges and they call for clarification of how to deal with them responsibly – individually or as a community.

Sports as a sphere of activity reflects a variety of challenges and interrogations posed by and for the whole of society. At times, these become even more pointed in the area of sports, as is the case, for example, with the range of issues surrounding doping or competitions. Hence it is little wonder that the specific discussion concerning forms of genetically engineered performance enhancement in competitive sports, and hence also gene doping, produces pertinent positions and lines of argumentation that also dominate the general ethical debate about enhancement of a healthy life. The spectrum of ethical judgments ranges from appeals to conservative values for a return to "clean sports" to liberal demands for lifting the prohibition on moleculobiological possibilities. While some diagnose a historically unprecedented corrosion of the value foundation of modern competitive sport and explicitly see its "nature" as threatened or perverted, others simply think that genetically engineered enhancement in elite sports legitimately catches up with already prevailing images and conceptions of the human being.

Condemnation and rejection of genetically engineered enhancements in sport relies in essence on two argumentative clusters: the protection of life and the ethos of sports. The critics view the protection of life as being especially endangered by potentially serious health hazards. These would be aggravated the more the techniques are applied without medical controls in the shadows of organized illegality.

With regard to the "ethos of sport," it is the lack of fairness that counts as one of the chief arguments against genetically engineered enhancement. Above all, particularly in the discourses on societal values, an additional important role is played by something that the WADA definition only hints at: they require and expect from competitive sport a "naturalness" meaning "authenticity" both of performance and of the athlete. In this way, the feasibility of examining the criterial function and argumentative heft of the nature concept also intensifies in step with genetically engineered enhancement in sport.

Ethics as a discipline prompts reflection on the values, norms, rules, and action guidelines of a society, of its subsystems and of every individual. With respect to genetically engineered enhancement, for example, questions that suggest themselves immediately are: What should the athlete (the coach, the physician) do or avoid doing for performance enhancement? What image of themselves do competitive athletes want to achieve? How can their decisions and actions be distinguished as either better or worse? May anything that is (technically) feasible be done to enhance performance? Should it be? To answer this kind of question arising from ethical reflection in the same breath also demands an accounting for the – ultimately arbitrary – dividing line between the realm of the permitted and the prohibited, such as between the legitimate and illegitimate alternatives for sports performance enhancement. To create more clarity here requires a sports ethic that evaluates model-based reasoning and arguments by proponents and opponents as well as thinking seminally about the goals and prerequisites of sport.

3.4 Law

Responsibility for the fight against doping, including the imposition of sanctions on doping infractions, is jointly discharged by sport and the state. The state's co-responsibility begins where the self-responsibility or autonomy of the sport bodies ends. Thus, sport's autonomy assumes concrete form in the National Anti-Doping Code (NADC), which defines doping as illegal and prohibits it. What substances and methods are banned is governed by the superposed stand-alone World Anti-Doping Code (WADC) prohibited list. The respective current version of this list, to which gene doping was added in 2003, is always a constituent part of the NADC. Under anti-doping regulations, the following violations are subject to sanctions on the sport-institutional level: use by an athlete, refusal to submit to testing, possession, trafficking and administration to another and other complicity. Basically, the athlete is accountable for all substances in his body ("strict liability").

The burden of proof rests with the anti-doping organization. The biggest challenge in this regard, particularly with respect to gene doping, is represented by what is called legally defensible detectability: for a violation to be provable requires the availability of an appropriate, reliable testing method. Both direct and indirect methods qualify for detection of possible gene doping. When it comes to legally defensible detectability, however, both types of methods are problematical. While the development and hence availability of direct methods is already critical from the biomedical perspective given the variety of moleculobiological approaches, an indirect method presents even greater problems in producing evidence since the method does not allow insight into exactly which banned substance or method was used. However, absent legally defensible detectability, it is also feasible to prove the use of banned substances or methods by other means, e.g., the athlete's admission, statements by witnesses or other evidence. Among NADC sanctions are temporary suspension of the athlete, ineligibility, disqualification of meet results, withholding of financial support, as well as imposition of financial sanctions. Under current law, this catalogue of measures can be transferred in its entirety to gene doping – subject to detectability.

The self-responsibility of the sport in combatting doping is complemented by the state's co-responsibility. Thus, for example, the state finances anti-doping measures and prosecutes doping-related bodily harm offenses, fraud, and drug law infractions. The legal basis for punishing doping infractions and, therefore, potential gene doping violations is found in the Criminal Code as well as the Drug Law, which incorporates the prohibited list. Third-party assisted doping is comprehensively regulated in these legal codes. Own-blood doping in principle is not subject to punishment by the state – imposing sanctions on it is therefore only possible at the level of the sport institution. In addition to the Criminal Code and Drug Law, the Embryo Protection Act could also come into play as national legal basis. However, this would require proof of the gene doping's effect on the human germline.

Chapter 4
Curricular tie-ins

The following table, by way of examples from the subjects of biology, ethics and norms, and sport, indicates possible curricular uses of the teaching unit in the state of Lower Saxony. Similar tie-ins can be found in other states, including ones related to law and social science. Within the cited disciplines also exist multiple possibilities for interdisciplinary or subject-combining integration of the teaching unit.

No.	Content-related scope	Page (nibis)
\multicolumn{3}{c}{Biology}		
	Source: http://db2.nibis.de/1db/cuvo/datei/kc_biologie_go_i_2009.pdf (as of 17.01.2013)	
	(The students…)	
1	(…) describe genes as DNA segments that contain information for production of gene products	
2	(…) using examples, explain the connection between genes and phenotype expression	15
3	(…) explain the effects of mutations on the phenotype	
4	(…) describe and explain biological facts by using suitable technical terms	
5	(…) develop questions about biological facts	18
6	(…) discuss complex biological problems whose solutions are controversial	
7	(…) evaluate possible short- and long-term regional and/or global consequences of individual and societal actions. Includes analysis of the factual and value levels of the issue as well as the development of courses of action	
8	(…) investigate complex problems and decisions in the light of social, spatial, and temporal pitfalls	19
9	(…) discuss opportunities and risks of transgenic organisms from the perspectives of different interest groups	

No.	Content-related scope	Page (nibis)
	Values and norms	
	Source: http://db2.nibis.de/1db/cuvo/datei/kc_wertenorm_go_i_12_11.pdf (as of 01.17.2013)	
Themes:		
10	Questions about what constitutes good conduct	
11	Basic positions of ethical argumentation	
12	Ethics in medicine and science	15
13	Questions about the nature of the human being	
14	Self-consciousness, free will, heteronomy	
15	Concrete action rules, binding expectations, intended and actually experienced ways of acting by individuals, groups, and societies.	16
16	Human dignity, human rights, human duties	
17	Inner conflicts: personal dilemmas	
18	Differentiating various forms of law and justice	
19	Examination of individual and societal problem areas of justice	
20	Law, guilt, and punishment	
21	Equal and unequal treatment	18
22	Ordering function of law, relationship between law and morality	
23	Diversity of interests and perspectives on societal, economic, cultural, worldview and religious levels	
24	Inequality of opportunity between parochial interests, lobbyism, and common interest	
25	Opportunities and dangers of scientific progress	
26	Ethical demands on scientists	26
27	Release of genetically modified organisms, genome research, genetics	
28	Human being as machine	31

No.	Content-related scope	Page (nibis)
	Sport Source: http://db2.nibis.de/1db/cuvo/datei/kc_sport_go_i_03-11.pdf (as of 17.01.2013)	
	(The students…)	
29	(…) make conscious decisions about their own athletic and physique development	19
30	(…) work systematically on acquiring, structuring, and using information, materials as well as media and apply the knowledge gained in differing contexts	20
31	(…) reflect differing assumptions regarding performance and interests, recognize those who are stronger and support or integrate those who are weaker	20
32	(…) possess knowledge, attitudes and values in relation to today's significant sports-related topics that cross subject boundaries, such as the problems of performance-enhancing substances in doping and problems of organized sports	21
33	(…) have the ability to think in a networked, multidisciplinary way, learn independently, form reasoned judgments and actions	21
34	(…) reflect on differing assumptions concerning performance and interests	
35	(…) distinguish among health-, fitness- and competition-oriented athletic pursuits.	41

Chapter 5
Options for using the teaching unit

The teaching unit can be used in the compact modular option (two class hours and up), the alternative option (two to three class hours), or the full classical option of five to seven hours. The modular option lends itself for when the time available is limited or individual dimensions and facets only are selectively taken up. In the classic option, all of the mentioned perspectives (law, ethics, and sciences) are considered and treated in in-depth. By contrast, the alternative option permits a reduced treatment of all perspectives using the teacher's information arranged by topic.

5.1 Modular options

Various possibilities are conceivable for the compact option. If the teacher does not want to go through the entire teaching unit, individual modular elements can be lifted from the teaching unit, such as the following:

5.1.1 Focus: Scientific foundations of gene doping

Here the teacher can concentrate exclusively on the scientific materials. The basic text can be used alone or in conjunction with the two in-depth texts on gene doping detection methods. This will take up between 90 and 135 minutes of teaching time. Available study materials for this include the basic text and the two in-depth texts.

Approach

It is recommended to begin with one of the two cartoons (p. 62, 63) and then have the students work on basic text individually or with a partner, followed by a joint evaluation of the results. The in-depth text on direct and indirect detection methods can, for example, be worked on using the partner puzzle. A teacher's evaluation rubric is furnished.

5.1.2 Focus: Legal foundations of gene doping

If the classroom work focuses only legal aspects of doping and gene doping, both of the basic legal texts can serve as resources. A double class period should suffice in terms of the time requirement. It takes about 25 minutes to read the material.

Next, the students should discuss the material once before the evaluation by the whole class. This should take up about 20 minutes. The remaining time is available for the evaluation.

Approach

A necessary starting point before using any of the legal materials is presenting the "The Marco Epowitz Case" scenario either by laptop/projector or else by passing out copies of the related documentation and discussing it. The legal questions and problems can then be discussed based on this case. Small group or pair work are suitable learning arrangements for working on the legal questions. It is a best practice to use the pre-structured worksheet. An approach without structuring help is recommended only for the most capable students. A teacher's evaluation rubric is furnished.

Alternative approach

The file of legal material on gene doping contains a PowerPoint presentation. It covers the contents of both legal material texts. This also offers the option of teaching the legal foundations frontally.

5.1.3 Focus: Ethical aspects of gene doping

Another option is to focus exclusively on the ethical questions bearing on the topic of gene doping. Here, the material covering the role of spectator/fan can be used in its entirety. K.H. Bette, A. Miah, T. Friedmann and F. Begov shed light on ethical and sociological questions surrounding this issue by addressing if and to what extent gene doping is acceptable. One of the two cartoons can serve as initial stimulus. The time requirement is two classroom hours. It takes about 20–25 minutes to read the material. Another 15 minutes are required for note taking and doing the work. The work is intended to help prepare for the discussion. A teacher's evaluation rubric is furnished.

Approach

A pro-con debate[1] can be held on the question "Is gene doping acceptable?"

The teacher should ensure during the preparatory work that a sufficient number of students adopt either a pro or con position. A vote can be taken before and

---

1 http://www.bpb.de/lernen/unterrichten/grafstat/46892/pro-contra-debatte (as of 1.18.2013).

after each debate. An evaluation of the debate performance and of the arguments presented (see evaluation rubric) should follow.

5.2 Alternative option

Another option is to provide an overview of the subject in two or three classroom periods. Here, use can be made of the text in the teacher's information arranged by topic. Either of the two cartoons can serve as initial stimulus.

This approach is best suited for working with high-performing learning groups because of the advanced-level textual requirements.

Approach

It is recommended to have the students explore the three perspectives by using the group puzzle.[2]

---

[2] http://lehrerfortbildung-bw.de/kompetenzen/projektkompetenz/methoden_a_z/gruppenpuzzle/(as of 1.18.2013).

5.3 Classical option

5.3.1 Teaching unit schematic overview

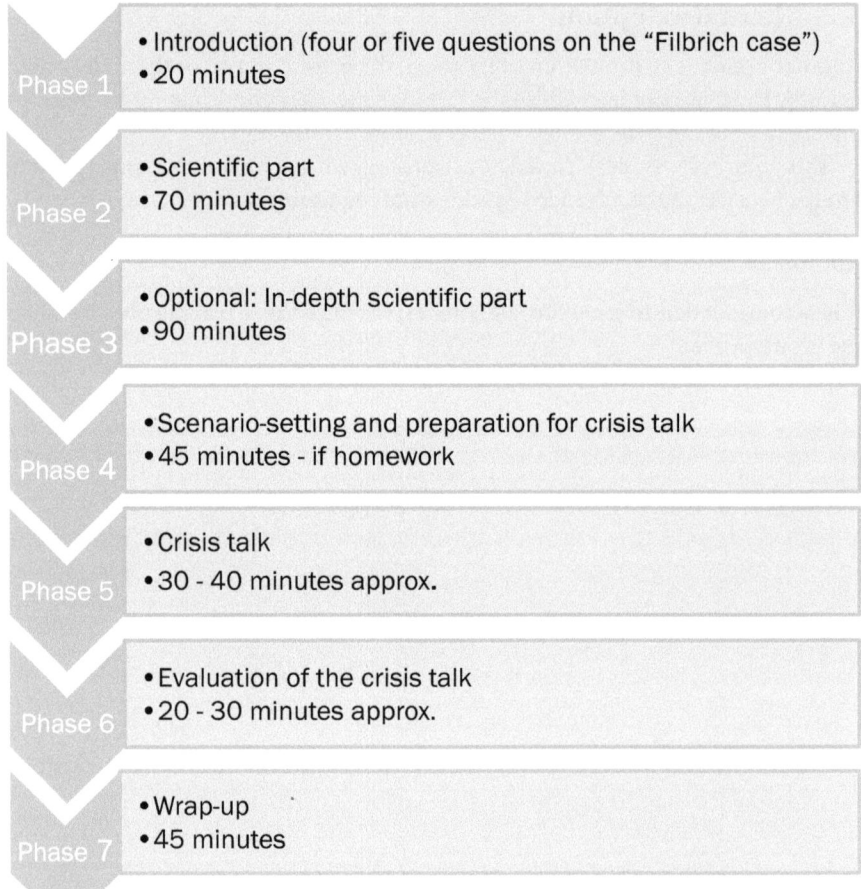

- **Phase 1**
  - Introduction (four or five questions on the "Filbrich case")
  - 20 minutes

- **Phase 2**
  - Scientific part
  - 70 minutes

- **Phase 3**
  - Optional: In-depth scientific part
  - 90 minutes

- **Phase 4**
  - Scenario-setting and preparation for crisis talk
  - 45 minutes - if homework

- **Phase 5**
  - Crisis talk
  - 30 - 40 minutes approx.

- **Phase 6**
  - Evaluation of the crisis talk
  - 20 - 30 minutes approx.

- **Phase 7**
  - Wrap-up
  - 45 minutes

Phases 3 and 4 (green) can be prepared in whole or in part as homework so that the classical option will take between five and seven hours of actual class time.

Phase 1 and Phase 2 are best done as a double period, as are Phases 3 and 4.

5.3.2 Course planning

Time (Min.)	Phase/Contents	Activities & Social Forms	Materials & Media
20	**Phase 1** Introduce and raise awareness of the issue	• Introduction and video clip • Have content fed back four or five questions; change positions in the room after each question and ask follow-up question after position change • Transition to scientific part: how can gene modification be achieved with technology?	• Laptop/projector + Filbrich file • Three "signs": YES/ NO/ UNDECIDED
		• Alternative start: cartoon	• Cartoon / overhead
70	**Phase 2** Scientific part	• Pair work (individual work) • Evaluation	• AB *Scientific foundations of gene doping* • Cartoons, illustrations, gene doping definitions as additional material
90	**Phase 3 (optional)** In-depth scientific part	• Repetition • Group work or • partner puzzle	• AB *In-depth 1 and 2* • Cartoons, diagrams, gene doping definitions as educational supplements **Continued on next page**

Time (Min.)	Phase/Contents	Activities & Social Forms	Materials & Media
10	**Phase 4** cenarioad preparation for crisisdssd talk	• Intro to the scenario Acting out the scenario/ news article • Question: what happened in Oberhof with what consequences? • Ask which actors are in the crisis talk and / or show slide with actors	• Laptop/projector + *News article* file or paper copy handout • Hand out materials for the five roles, also 5 name tags and 5 felt- tipped markers
		• Announcement that 45 minutes are allocated for preparation • Five groups for five roles • Allow room for the crisis talk	
40	**Preparation phase**		See "Phase 4" above
5	**Direct transition to crisis talk**	• Discussion participants go to the *podium* • Audience members form a semi-circle	• Audience assignments
30	**Phase 5** Crisis talk		• Name tags • Audience assignments
20–30	**Phase 6** Evaluation discussion: a. Mental states b. Argumentation c. Legal questions	• Plenum discussion	• Audience assignments
(+30)	**Conclusion** a. Actor constellation b. Biographical case and personal experience c. Evaluation questionnaire (30 minutes)	• Plenum discussion • Individual work	• Blackboard

Light gray means: can be done entirely or partially as homework as desired.

5.3.3 Detailed guidelines for the individual phases

Phase 1: Introduce and raise awareness of the issue

Using the video clips,[3] the students can be prepped and activated for basic problems of the topic and also ones recurring throughout the teaching unit.

The teacher can open the teaching unit with a few words of introduction, such as:

"Jens Filbrich is a competitive athlete. He has been mentioned as an Olympic hopeful. He did not manipulate his body, has not done anything illegal and has never been injured. But it looked like his career was about to end anyway. After you watch this one minute introductory clip, please be prepared to tell me what big problem is dogging him."

After the students have briefly sketched the problem in the Filbrich case, the teacher prompts the students to imagine themselves competing against Jens Filbrich and then take a stand on the following questions.

The teacher next goes down the below list of questions one by one and has the students each line up behind one of the sheets (see Materials section) marked YES, UNDECIDED, and NO that are placed on the table or on the floor. The teacher then asks why the students lined up the way they did and not another way. At this point, a discussion among the students can also get started. The results of the positioning can be recorded on the relevant transparency and discussed if necessary.

1. Is this a fair competition?
2. Should the athlete be excluded from the competition?
3. Do you like taking part in this competition?
4. Should there be a separate performance category for athletes that have a natural advantage? Or: Should there be separate performance categories for athletes that have biochemically manipulated their bodies?
5. Does it make a difference if the gene defect is inherited or the gene was modified biochemically?

After the last question, you can transition to the following scientific part by asking, for example, how a gene can be biochemically modified. Normally, the response to what are likely to be vague answers can be that the exact ways of biochemically engineering genes will be dealt with in the remainder of the course. Alternatively,

---

3   http://www.youtube.com/watch?v=zIylUGqacNw (accessed on 18.01.2013).
    The file can also be ordered on a CD (in HD) by email containing a mailing address. Please send the email to: faul@dshs-koeln.de. The CD also contains the three minute news segment and the PowerPoint file titled: Legal basics of gene doping.

in place of the question about possibilities of genetic modification – after questions four or five – the transition to the scientific part could be effected with the help of either cartoon (see Materials section).

This introduction during the double class period will take an estimated 20 minutes.

Experience shows that the students will respond to this introductory phase with interest and in an engaged manner. The rather unusual getting up out of their seats may also contribute to this and induce a favorable dynamic as you go through the rest of the unit.

Phase 2: Scientific part

The teacher can now have the students start on the work sheets dealing with the scientific foundations of gene doping. 70 minutes are set aside for the basic part (MATERIAL 1) of the scientific information (i.e., without the two in-depth texts).

Having the text and the exercises worked on in pairs has proven itself in practice. Allocate 35 minutes for option one. Use the remaining 35 minutes for discussing the exercises.

If needed, you can bring the diagrams and cartoons in from the Materials section as additional material. For example, the definition of gene doping can be examined in depth: a.) Give the gist of the definition; b.) Point out the change in the definition and ask what explains this change; c.) Even scientific definitions are not carved in stone and reflect political / societal decisions.

Phase 3 (optional): In-depth scientific part

Besides the basic information on the subject of gene doping, potential direct and indirect detection methods (MATERIAL 2 and MATERIAL 3) can be covered in-depth during an additional double class period. Allocate 25 minutes to work on this.

Option 1
The texts can be worked on in small groups and can then be evaluated frontally.

Option 2
A partner puzzle can also be deployed in proficient groups. In that case, the class period might look something like this:

1. Review and introduction
2. a.) Half the students (A) work on "Direct detection methods."
    b.) The other half of the students (B) work on "Indirect detection methods."

3. Next, each student A makes a presentation of own work done on the exercises while student B takes notes on A's expositions In reverse, student B makes a presentation to student A. Now everyone is knowledgeable about "everything."
4. Finally, the results can be reinforced for the entire class, ideally by having a student who only had the exercises presented and did not work on them introduce and explain the solution.

Additional materials: See Phase 2.

Phase 4: Scenario (news article) and preparing for the crisis talk

Before reading or playing[4] the news report (MATERIAL 4) that furnishes the basis for the following crisis talk, do a brief lead-in to the article. Example: "Jens Filbrich, the cross country skier, made the headlines through no fault of his own. Another, much better known athlete, Marco Epowitz, has set off a media storm."

Why? We'll find out in the two-minute news report. Sum up what happened in Oberhof and what should happen next."

Transition straight into the preparatory work after the summary. It might be opportune at this point to ask who would take part in such a crisis talk. This does not call for a discussion, rather it is designed simply to move expeditiously into the crisis talk's preparatory phase.

At this time the teacher hands out the materials for the individual roles (MATERIAL 5–25, volume 2). Each of the five groups should be made up of three to five students. All groups will be ready to go after 40–45 minutes. Of this, about 20–25 minutes are devoted to reading the texts and the remaining time goes to internal discussion and note taking. In practice, thinking of a name has proven to be highly motivating and useful for the flow of the discussion (see Materials section). The students can let their imagination roam by thinking up names for coaches like "Ana Bolics" or "Miles Tugo" that fit the scenario. The materials used to prepare for the crisis talk were thoroughly tested so that a very high independent work quotient is feasible. Experience has shown that the teacher's assistance is most likely to be called for in the lawyer group. The teacher

---

[4] The news report can either be shown as a three-minute video, or else be handed out to the students as a room document. In the first instance, you can download the news report from the website at http://www.vidup.de/v/ktKff/ or request it in file format by an email that includes your mailing address to faul@dshs-koeln.de. You will receive a free CD containing the three-minute news report and the one-minute clip "Filbrich" segue in HD.

must be sure to point out that the coach should back up Epowitz one hundred percent. Also make clear to the athlete group that Epowitz has to deliberately justify his gene doping. Each group should be allowed to decide who will take on which role in the crisis talk, since only one from each group can be the discussant for the podium discussion, with the others being audience members with opportunities to ask questions.

After the preparation time for the crisis talk is up, organizing pointers are given before the start of the crisis talk (Audience Tasks, MATERIAL 26, volume 2). The organizing directions and the start of the crisis talk must be kept strictly separate. For this reason, it is recommended that a student give the organizational directions so that the crisis talk moderator does not have to.

The talk itself should be moderated by the teacher or by a high-performing student. Detailed pointers for the moderation follow on the next few pages.

Options

Preparing for the crisis talk can also be assigned as homework. In that case, show the scenario at the conclusion of the scientific part.

Phase 5: Crisis talk

Sketch of the discussion start and topic areas

The aim of this somewhat highly planned way of proceeding is to provide a practical way of getting the crisis talk started. The following illustration is only meant as a suggestion; however, it is has repeatedly proven itself in practice.

Moderator's introduction

Welcome to *"Now what? Will it be all Epo or what? – After the Epowitz scandal is genuine sport finished?"*

I want to warmly welcome everyone here in the studio and those watching at home to today's broadcast. My name is _____ and I'm pleased to have a chance to take up this fascinating topic with our guests today.

For those viewers who just tuned in: We are focusing on an incident that occurred during the biathlon at the World Cup in Oberhof. Both A and B tests allegedly showed that Marco Epowitz manipulated his body. We have come together here to shed more light on this incident with a crisis talk. Since this is a red-hot happening the public also has something to say about it. You will have several opportunities to ask questions. Despite the emotionally charged mood, please allow everyone participating to make their case without interruption. We have set a 30 minute time limit for the talk. At its conclusion, we will show an interview about the events at Oberhof with Mr. __, vice president of the International Olympic Committee. Now, let me briefly introduce our guests:

(**Optional:** The discussants do their own introductions.)

The crisis talk participants

- **Marco Epowitz**, you are considered an exceptional talent, have an extraordinary amount of national and international experience and successes in the biathlon and now are facing accusations of gene doping.
- **XX**, you have been coaching Marco Epowitz for the past five years. They say that you laid the foundations for his success through unconventional and effective methods.
- **XY**, you are the president of the German Ski Federation. Your federation is the umbrella organization for German ski associations. You represent the interests of German ski sport.
- **XZ**, you are a long-time biathlon fan, especially of Marco Epowitz. As you said during the introductory talk, you don't hesitate to travel to other continents to follow the sport and your hero live.
- **YZ**, you are a well-known specialist in sports law and familiar to many as an expert from appearances on other shows about sports law.

It's great to have YOU all here!

Our thanks to the one most affected by all this for joining us today. How do you feel after the media hurricane, Mr. Epowitz?

- Epowitz makes a statement.
- How do the others present feel about the event or the accusations?
- The sentiments are one side of the coin, statutory and legal considerations are
- the other. With us is an expert to brief us on them. (For answers to the questions, see the "lawyers" file.)
- Before we get to you as our (legal) expert, Mr. YZ. How do you gentlemen as coach and federation official respectively see it?
- [Legal question 1] "Will Marco Epowitz have to go to jail, Mr. YZ?"
- The answers given by the federation official and the trainer "…"
- "Thank you for sharing that. Now, let us hear how the expert sees it."
- The lawyer answers: "…"
- [Legal question 2] "This leads to another question: Did Epowitz defraud anyone (Section 263, Criminal Code) by having gene doped himself? As a fan, do you feel cheated?"
- The fan answers: "…."

- "How do you as a lawyer see whether Marco Epowitz cheated anyone?"
- Following these two questions about legality that also involve the other discussants, the floor is opened for the discussion, which, at first, steers away from legal questions. The next two legal question will be saved for a later point in time.
- [Legal question 3] Are others involved subject to criminal prosecution if they gene dope Epowitz? So, for example, his coach, if he provided or injected the substance?
- [Legal question 4] What should Marco Epowitz expect in terms of sport legal sanctions? What do NADA and WADA provide for in such cases?

The content of the discussion can be expanded, for example, by asking whether gene doping has the potential to level the playing field for athletes with disabilities.

- More questions can be generated from the suggested questions (see below) or they could be formulated situationally by going in-depth on student answers. In addition, a few opportunities for questions from the audience should be provided.
- The discussion ends with a final round in which everyone answers a question written on the blackboard, such as: Is it all over for true sport? Or a more concrete concluding question: Taking all aspects into account, how should Marco Epowitz be dealt with?
- We'll stay on top of the Epowitz case. Tune in tomorrow evening at seven o'clock. But now, let's go to our live interview with the vice president of the International Olympic Committee

Alternative start: Skip the moderator's introductory round (i.e. the discussants introduce themselves) and instead start off with a question for the audience:

- You are all ski enthusiasts, otherwise you would not be here today. Epowitz is said to have manipulated his body illegally. How do you see it? (gather opinions). And/or the question: Should expulsion of Epowitz from the world cup be rejected?
- Thank you very much! Let us find out how our guests view the incident.
- The discussants introduce themselves.
- We are fortunate in having the man himself here with us today. How do you feel after the media storm?

Possible questions addressed to the discussants by the moderator

Questions for the lawyer
- Could you elaborate which legal considerations come into play in cases of gene doping?
- What does it take to substantiate a finding of cheating?
- Could someone be subject to prosecution for causing physical injury?

Questions for the athlete
- Do you think your being banned from competition is just(ified)?
- Why did you engage in gene doping?
- When did you decide to resort to gene doping? Did anyone influence your decision?

Questions for the coach
- What is your position on gene doping? Are you for or against it?
- Were you previously aware of your athlete's doping practices?
- Would you advise your athlete in the future to engage in gene doping?
- Was it fair play to have compensated for your athlete's natural shortcomings with gene doping?
- Do you see a connection between gene doping, competitive sports and a meritocratic society?

Questions for the spectator
- How do you feel about gene doping?
- What do you see as the pros and cons of gene doping?
- What consequences would legalizing gene doping have for you and for the sport in general?
- Is gene doping morally repugnant?
- What would your reaction be if gene doped athletes could compete in their own super performance category?
- Do you see a connection between technological innovations in other areas of society (Internet, Facebook, "brain doping") and gene doping in competitive sports?

Question for the federation official
- What values and goals play key roles in elite sports?
- How should we regard gene doping in this context?

- Isn't it evident that athletes have to (gene) dope in order to meet the performance expectations for elite sports (higher, faster, farther, the record-setting logic)?
- How will you deal with Epowitz now? What is the significance of the Epowitz case for the future of your federation?

More suggested crisis talk questions

The questions below should be answered / discussed by the students with reference to their work on the material and the assignment but also *ad hoc* by "connecting the dots."

The competitive sport system
- What is sport? What is competitive sport?
- Do we need to devise new criteria for competitive sports and regard gene doping as a normal next step?
- Does the competitive sport system inevitably lead to gene doping? Should you be able to "opt in" to gene doping?
- What roles do coaches play in particular?

Health
- What about athlete health?
- What health risks does an athlete run who, knowingly or not, is gene doped?
- Is a gold medal worth putting your health on the line? (goldman dilemma)

Justness
- Would it be more just to allow gene doping for all?
- What role does justness even play in competitive sports?
- Is it just that some athletes have innate genetic advantages when it comes to endurance and strength? Why/why not?

Society and public
- Will gene doping change how the public thinks of sports?
- Are there society-wide risks?
- Is a sport that has no records without gene doping still attractive?
- In the stands, it is not about ethical consciousness. The crowd enjoys the action and the outcome above all. Without record results, sport would become boring. Are we ready for that?

Naturalness / artificiality
- What divides the artificial from the natural? Are competitive athletes natural? Is gene doping not altogether naturally part of it?
- To what extent does gene doping contribute to the estrangement of the athlete from his body?
- Is gene doping just another, logical step in the ongoing convergence of the human and machine/technology?
- Is there a difference between natural, inherited and artificially doped gene mutations? In what degree?
- To what extent could practices of gene doping influence the athlete's daily life (i.e., life outside sport)?

The law
- Who can be prosecuted for what in cases of gene doping?
  Punishable offenses include trafficking, prescribing, and using drugs on others for purposes of sport doping under paragraph 6a, 95 Medicinal Products Act (MPA = Drug Law). Further, gene doping "of others" can be considered to cause physical harm under Section 223 of the Criminal Code. Even if the athlete permits it, it does not exculpate the third party if doping risks a serious bodily injury and so makes it an illegal act (Section 228, Criminal Code). A coach or other third party becomes culpable when providing support to someone else in committing these acts, i.e., helping them (aiding) or urging them to do so (abetting).
- To what extent can the gene doped athlete be guilty of an offense?
- The gene doped athlete can be personally prosecuted under Section 263 of the Criminal Code for defrauding the organizer or his sponsor. If found in possession of more than a minor amount of doping agents, Sub-section 6a, para. 2a may also apply.
- On the level of sports law, the gene doped athlete has violated the NADA/WADA code so that he also faces legal sanctions under the sports / federation statutes. To what degree does this apply in the Marco Epowitz case?
- What important statutory and legal principles apply in relation to the Epowitz case and gene doping?

Fairness
- Is gene doping reconcilable with the values of the sports federations / competitive sports?

- What is the relative importance of so-called values (like fairness) espoused by the sports federations (for the individual athlete, the federation itself, and others)?
- Should there be special performance categories for gene doped athletes? For example, "mutant" categories"?
- How fair is it to compensate for these "natural deficiencies" with gene doping?

Level playing field

- Could gene doping equalize genetic differences between competitors in a competition in the sense of leveling the playing field?
- What should equality of opportunity relate to? To rules that apply to everyone equally or also to the opportunity to create equal physical preconditions?
- Equal opportunity, level playing field – is this even realistic? Is it desirable?

Phase 6: Appraising the crisis talk

Appraisal discussion

Once the students have returned to their seats after the crisis talk and have stepped out of their roles, the reflection and metacommunication phase can begin. The following reflection levels might be considered depending on the class situation.

a. To help get the evaluation off to a good start, we recommend asking the students how they approached their roles.
b. As a next step, the audience members can refer to their notes (audience assignments) to raise questions like the following:
    - Which discussants favored acceptance of gene doping?
    - How did the role players argue? (Possibly record this on the blackboard – see evaluation rubrics). Which arguments did they cite?
    - Which arguments were missed?
    - Give reasons why a particular role holds more appeal.

Lastly, the two legal questions should be cleared up with reference to the audience assignments.

Supplemental questions

What significance do you think gene doping will have for sports in the future? Does gene doping align with a contemporary society that views itself as a performance society?

Phase 7: Conclusion

The teaching unit's last class period can be devoted, for example, to a thoroughgoing reflection in writing about the teaching unit (see the critique sheet in Materials). Alternatively, a mix of verbal summation and use of the critique questionnaire can be practiced.

Approach

The teacher prompts: "Name actors/parties that were directly or indirectly involved in the Epowitz case. There are more than could take part in the crisis talk." The teacher writes down the names called out (refer to Figure 1 for an example).

The actor constellation is to be worked through explicitly one more time. The arrows are intended as visual reminders of the pressure that stakeholders put on the athlete. As part of this exercise, also ask:

- What interests do the individual actors pursue?
- Which interests clash?

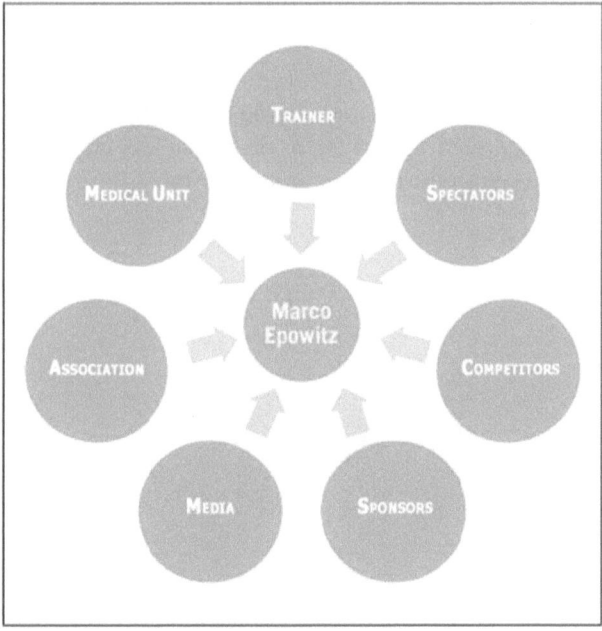

Fig 1: Parties to the doping scandal
Source: the authors

This makes explicit the complex web of relationships and interests within which the athlete must move.

- Within this constellation, how could Marco Epowitz manage to do gene doping?

On the one hand, the large arrows symbolically convey an impression of the possible pressures weighing on Marco Epowitz. On the other hand, the teacher can also point to the so-called "biographical trap."

The expression "biographical trap"[5] conveys how, in the course of an athlete's life, competitive sport makes growing demands on his or her time, social life and finances, so that abandoning the "sport system" is hardly feasible at that point. In line with this logic, a man like Marco Epowitz must do his utmost to achieve success, with the now familiar consequences.

A possible question: What can the athlete do to handle this?

After this abstract pigeonholing, a bridge can also be built to the students' own life.

- What experiences have you had with regard to the constellation of actors?
- Do you view gene doping as an acceptable way of dealing with the "biographical trap"?

With this question, the transition to the critique questionnaire (MATERIAL 27) has also been accomplished, so that the students now can turn to completing the questionnaire. This brings the teaching unit to a close.

---

5   See Bette et al., 2006, p. 135.

Chapter 6
Additional readings

Beiter, T. & Velders, M. (2012). Pimp my genes – Gendoping zwischen Fakten und Fiktionen [*Gene doping between fact and fiction*]. Deutsche Zeitschrift für Sportmedizin, 63, 121–131.

Bette, K.-H. & Schimank, U. (2006). Doping im Hochleistungssport. Anpassung durch Abweichung (Erw. Neuauflge) [*Doping in high performance sports. Adaptation through deviation (expanded and revised ed.)*], Frankfurt a.M.: Suhrkamp.

Bette, K.-H. & Schimank, U. (2006). Die Dopingfalle. Soziologische Betrachtungen, [*The doping trap. Sociological observations*] Bielefeld: transcript.

Franke, E (Hrsg.). (*in press*). Herausforderung Gen-Doping. Bedingungen einer noch nicht geführten Debatte [*The gene doping challenge: requirements for a debate that has yet to happen*], Bielefeld: transcript.

Gerlinger, K., Petermann, Th. & Sauter, A. (2008). Gendoping – Wissenschaftliche Grundlagen, Einfallstore und Kontrolle [*Gene doping – scientific foundations, gateways, and control*] (Studien des Büros für Technikfolgenabschätzung beim Deutschen Bundestag, Bd. 28) [(*Studies of the Office of Technology Assessment, German Parliament, vol. 28*)]. Berlin: edition sigma.

Körner, S. & Schardien, S. (Hrsg.) (2012). Höher – schneller – weiter. Gentechnologisches Enhancement im Spitzensport [*Higher – faster – farther. Genetically engineered enhancement in elite sports*], Münster: Mentis.

Internet links

http://www.bpb.de/lernen/unterrichten/grafstat/46892/pro-contra-debatte (accessed on January 17, 2013).

http://lehrerfortbildung-bw.de/kompetenzen/projektkompetenz/methoden_a_z/gruppenpuzzle/ (accessed on January 17, 2013)

http://www.vidup.de/v/ktKff/ (accessed on January 17, 2013).

http://www.youtube.com/watch?v=zIylUGqacNw (accessed on January 17, 2013).

http://www.gentechnologie-im-sport.de (accessed on January 17, 2013) (Note: Here you can also find a selection of articles, audio and video resources dealing with the subject of gene doping for use in the course.).

Chapter 7
Annex

7.1 Material for Phase 1

Lead-in for the four or five questions (up to 1:12 minutes)

1. http://www.youtube.com/watch?v=zIylUGqacNw (accessed on January 18, 2013).
2. Alternatively, the file can also be ordered on CD (HD quality) by email to: Faul@dshs-koeln.de stating the delivery address.

The CD also contains the three-minute long news article and the PowerPoint file: *Legal foundations of gene doping.*

Question catalog – barometer method

1. Is this a fair competition?
2. Should the athlete be disqualified from the competition?
3. Would you want to take part in this competition?
4. Should athletes with natural advantages have their own performance category?

   – or –

   Should athletes who have manipulated their bodies biochemically have their own performance category?
5. Does it make a difference if the genetic defect is inborn or the gene is modified biochemically?

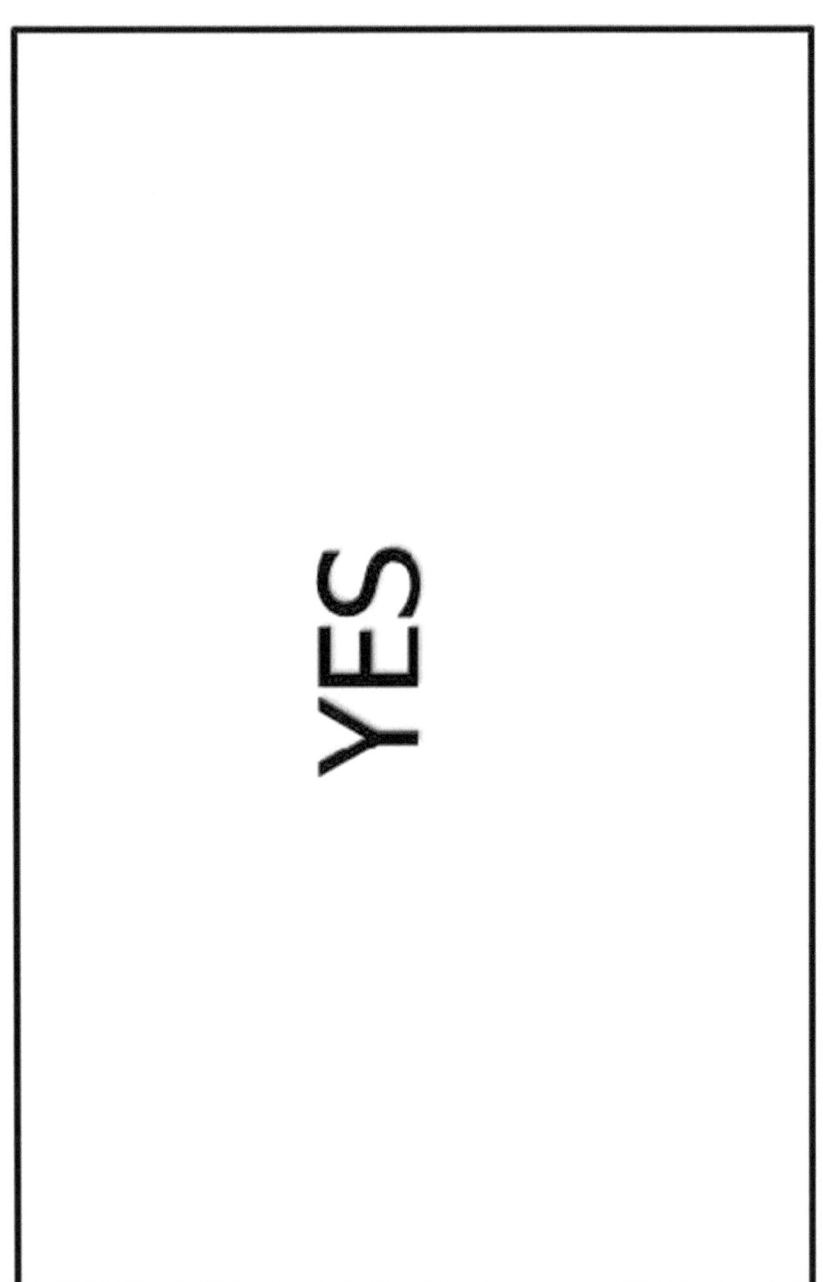

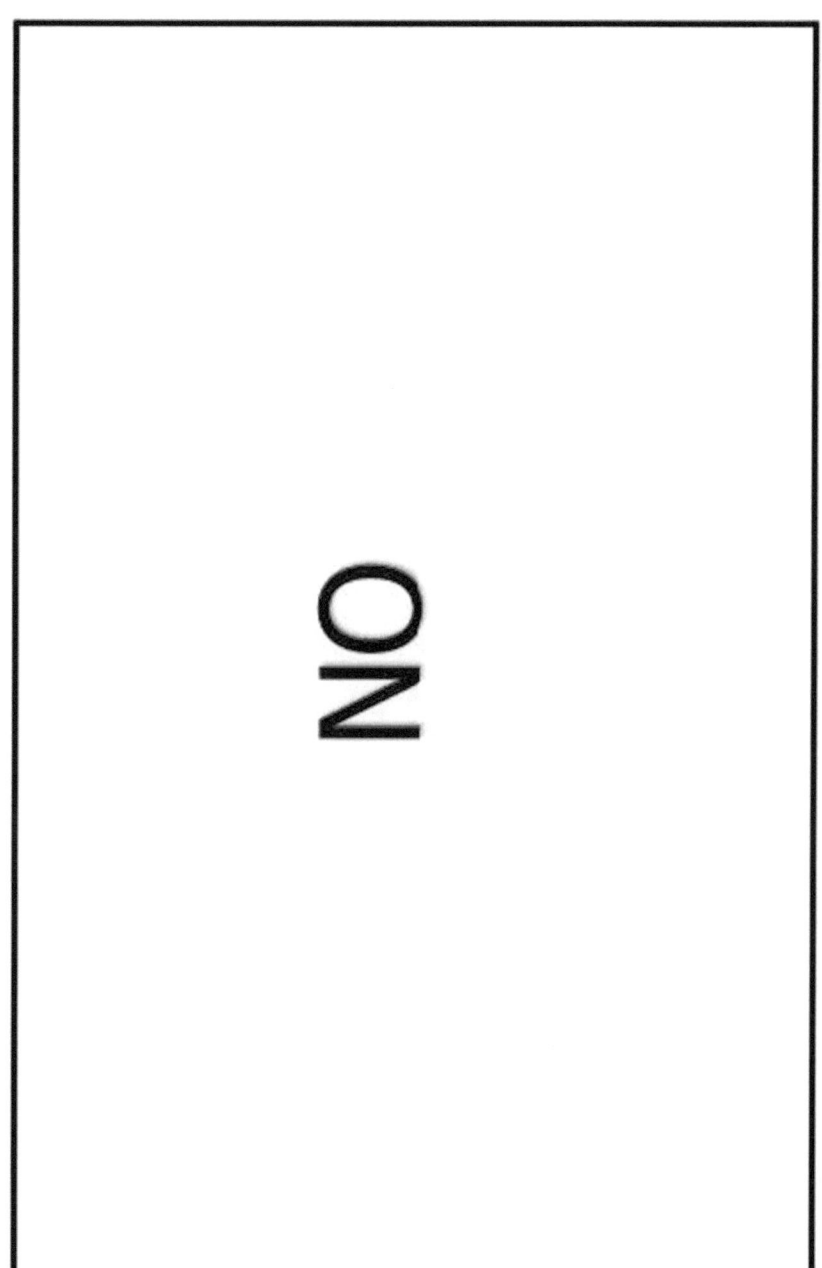

UNDECIDED

1. Is this a fair competition?

| YES: | NO: | UNDECIDED: |

2. Should the athlete be disqualified from the competition?

| YES: | NO: | UNDECIDED: |

3. Would you want to take part in this competition?

| YES: | NO: | UNDECIDED: |

4. Should athletes who have natural advantages have their own performance category?

| YES: | NO: | UNDECIDED: |

5. Should athletes who have manipulated their bodies biochemically have their own performance category?

| YES: | NO: | UNDECIDED: |

6. Does it make a difference if the genetic defect is inborn or the gene is modified biochemically?

| YES: | NO: | UNDECIDED: |

7.2 Material for Phase 2

Scientific basis of gene doping

Definition of gene doping

The World Anti-Doping Agency (WADA) defines gene doping as the transfer of nucleic acids or nucleic acid sequences (DNA, RNA) as well as normal or genetically modified cells with the potential to enhance sport performance.

Nucleic acids

DNA is double-stranded and is packed into chromosomes in the nucleus of every cell in our bodies. RNA is produced in the nucleus as a single-stranded copy of DNA; it exits the nucleus and is used in the cytoplasm as pattern for the synthesis of proteins (gene expression).

While a living organism's cells contain mostly identical DNA, different cells produce differing RNAs and proteins depending on their function (skin, muscle, liver cells, etc.) (Differential gene expression).

Mutations

 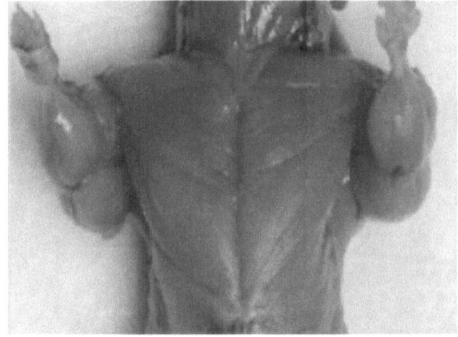

Fig. 1: Genetically modified mouse.
Source: Lee 2007; PLoS One. 29:2(8):e789

Fig. 2: Genetically modified mouse.
Source: Lee 2007, PLoS One. 29;2(8):e789

Mutations in most cases are naturally occurring random changes in the DNA. From mutations it is possible to tell how large the effects can be when a living organism is modified genetically. For example, in Belgian blue cattle, the gene for myostatin, a hormone that inhibits muscle growth, is defective due to a mutation. This causes the animals to grow bigger muscles. In experiments on mice, this effect has already been successfully produced through genetic engineering (Figs. 3, 4).

Gene doping vs. conventional doping

The concept of gene doping is based on the principles of gene therapy. In contrast to conventional doping, gene doping is designed to get the athlete's body to the point where it produces the doping substance (e.g., EPO) itself. Under certain circumstances, after the gene transfer, the body may produce the doping substance for the rest of its life, and, in the worst case, this happens in an uncontrolled manner and leads to severe side effects. There are to date few longitudinal studies on this topic, and the side effects naturally are also highly dependent on the kind of infiltrated gene and its regulation.

Germline therapy vs. somatic gene therapy

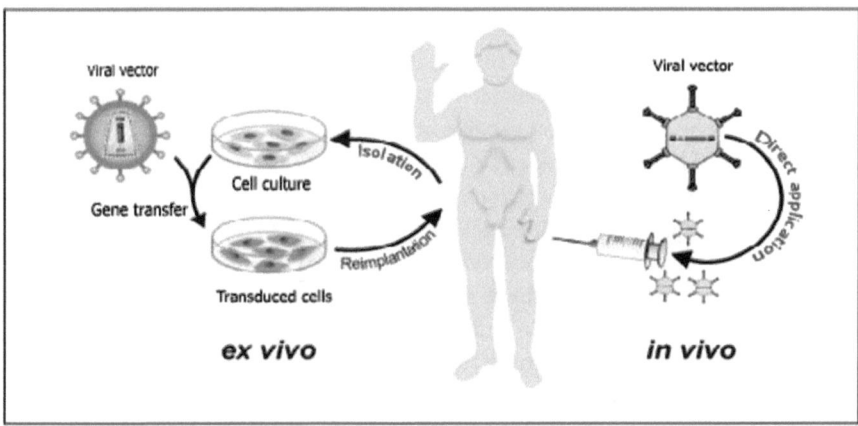

Fig 3: Options for gene transfers with somatic gene therapy.
Source: Beiter & Velders (2012) DZS. Jg. 63, Nr. 5. p. 123

Gene therapy differentiates between germline and somatic therapy. In germline therapy, germline cells (egg or sperm) are modified genetically. This type of gene therapy can be inherited by the offspring. In somatic gene therapy, (from "soma," the Greek word for body) as the name implies, the body's non-reproductive cells are manipulated genetically – as a rule, this form of gene therapy is not inherited by the next generation. In somatic therapy, the gene transfer with the therapeutic gene (also called a transgene, e.g., EPO) takes place under controlled conditions outside the body (ex vivo; see Fig. 6). This type of gene transfer has the advantage that only cells taken from the body are manipulated genetically. In contrast, with in vivo gene therapy the gene transfer occurs inside the body, for example, by injection into the bloodstream or the musculature. The in vivo form of gene therapy may be easier to implement technically than ex vivo therapy, but the side effects of in vivo therapy can be much more severe as the result of a massive immune reaction against parts of the viral vector. In the worst case, it can lead to death through multiple organ failure.

In order to get the therapeutic gene into a patient's body, gene therapy makes use of the ability of viruses to penetrate the cell membrane efficiently and, in some cases, even into the cell nucleus. In this respect, viruses serve as "gene taxis," or in technical language "viral vectors."

Which body cells are particularly suited for gene transfer in gene doping?

This of course depends on which aspect of physical performance capability is to be enhanced. It is possible to increase muscle strength by genetically manipulating muscle cells. However, an in vivo injection, perhaps in the thigh, will only suffice to reach a small part of the superficial muscle fibers. After embryonic development stops, muscle cells are no longer capable of dividing. This offers the advantage that genes infiltrated by gene therapeutic means into the cells remain there if they do not integrate themselves in the chromosomes (episomal maintenance).

However, this property of the musculature represents an inhibiting factor for gene therapy, for example, when viruses are employed that will only integrate with actively dividing cells. Still, in the musculature there are stem cells (satellite cells) that do have the ability to divide and foster muscle growth with their DNA to regenerate muscle after hard training. Only future research projects can show if it is possible to modify these cells genetically in order to boost performance. Since muscles make up about 30–40% of body mass, this tissue is also suited for production of non-muscular gene products, such as EPO. In this case, using an appropriate virus, the EPO gene will be infiltrated into the musculature where it will produce the EPO protein. The EPO will enter the bloodstream as a growth factor to stimulate production of red blood cells (erythrocytes) that will then improve the blood's oxygen transport capacity and hence endurance sport performance.

Transferability of animal experimental findings to humans

Gene therapy has already achieved remarkable results in animal experiments. Transferring them to humans, however, so far only succeeded in a few isolated instances. In animal experiment it has even been shown that transgene activity can be turned on or off as desired by administering drugs (on/off system). Some of the gene transfer studies of humans found unexpected, occasionally grave side effects, such as the emergence of blood cancers (leukemia) or multiple organ failure stemming from severe immune reactions. Before gene therapy can be applied safely more intensive research and development will be needed.

Detection of gene doping

Following successful gene therapy, the body will produce the transgene's gene product (protein) and thereby make it impossible for conventional doping detection methods that are based on detecting artificially produced doping substances to detect the protein. For this reason, gene doping detection must deploy a new method. Various moleculobiological techniques offer themselves in this regard.

Assignments

1. Describe the differences between gene doping and conventional doping.
2. Explain possible methods for transferring genes used in gene therapy.
3. Weigh the potential opportunities and risks of gene doping. As you do so, think carefully about what yardsticks to apply for your evaluation.
4. What are the special properties of muscle cells and how do they affect treatment by gene therapy?

Evaluation rubric

For question 1

- Gene doping = one-time treatment/injection (long-lasting, possibly lifelong gene production) vs. regular treatments/intake with conventional doping.
- Self-production vs. giving of an artificially produced doping substance.
- Bio-identical vs. chemically-modified substance (recombinant).
- Substances produced in gene doping cannot be detected.

For question 2

In vivo

- Pro: "practicability," low effort, just one injection, etc.
- Con: uncontrolled spread of viruses and integration of the transgene in various (possibly undesirable) cell types, side effects, etc.

Ex vivo

- Pro: controlled gene transfer
- Con: costly, technical procedure, biopsy and tissue reimplantation, not suited for all cell types

For question 3
- Regular treatments unnecessary, possibly more cost-effective, and highly effective
- Physical changes lasting a lifetime ≠ doping substances degraded after week/months, side effects, research gap (e.g. leukemia), etc.

For question 4
- Muscle cells in mature organisms are postmitotic, i.e., no longer capable of dividing. This is advantageous when a transgene that is introduced therapeutically into a muscle cell gene fails to integrate chromosomally but instead remains "episomally" intact in the cell nucleus. If the cells were mitotic, the transgene would be lost during cell division.
- In non-dividing cells, the transgene thus is preserved in the nucleus and under the right conditions is able to manufacture the related protein product indefinitely.
- Musculature makes up approximately 30–40% of the total body mass. For this reason, this tissue type is highly suited as a production facility for non-muscular gene products, such as EPO, that are then released into the bloodstream to exert their effect there.

7.3 Material for Phase 3 (optional): In-depth scientific section
In-depth text 1: Indirect detection of gene doping

With indirect detection methods, it is not the doping substance to be identified that is detected by the test (unlike direct methods) but rather its effects on certain blood values. So, for example, when doping with testosterone is suspected, it is the ratio of testosterone and epitestosterone in the body that is analyzed. Another example is indirect detection of EPO. This is usually done using the hematocrit (volume percentage of red blood cells in blood), the hemoglobin concentration, the number of reticulocytes (immature erythrocytes), macrocytes and a few other parameters. The standard values differ from individual to individual, even if in medical practice certain average values are specified as being normal.

This can result in a hematocrit value that is read as elevated actually being normal for some people. In this connection, it needs to be asked if it is not simply those naturally caused deviations from standard values that in many cases account for athletic talent.

This requires taking into consideration the differences between individual athletes. Individual blood values observed over a longer period are therefore more meaningful than one-off blood tests. To capture such values is the purpose of setting up blood profiles. This involves taking blood samples from the athlete at regular intervals and under varying circumstances, for instance before and after training camps, with the results then entered into a so-called blood passport. This makes it possible to flag deviations from the individual value reference range that may indicate doping.

However, for the numerous gene doping candidate genes such standard values are not yet known and great deal of study will be required to understand the molecular interrelations and influencing factors on each of the genes.

A precondition for effective doping detection using indirect testing methods is that all candidate genes and their molecular signaling pathways must be known. As part of indirect detection methods, blood proteins will ultimately be identified that play a role in the regulation of each gene's physiological activity. For example, in cases of manipulating the myostatin gene, changes in the concentration of myostatin inhibiting proteins (e.g., follistatin) might be considered as indirect evidence of doping. Still, it remains a challenge for anti-doping research to even identify suitable blood proteins for doping detection, because these parameters must invariably only be modified by doping and not by physiological factors, such as physical activity. For this, scientists must determine what concentration is normal to begin with. In determining specific maximum limits, they must establish clearly to what extent the protein concentration can be altered naturally, as for example by training, endogenous hormones or certain foods, etc. It follows that intensive research will still be required in the future to achieve the ability to detect doping in general and gene doping in particular.

Assignments for the scientific in-depth text

1. Name the differences between direct and indirect doping detection methods.
2. Form an opinion on the challenges posed by direct doping detection. In doing so, keep in mind the characteristics that the particular blood parameters must exhibit.

Evaluation rubric

For question 1

Direct doping detection: detection of synthetic doping substances; substance used in doping → directly detected in the sample.

Indirect doping detection: the effects of the doping substance / gene modification on specific blood values are analyzed. Detection takes place based on changes in the concentration of certain blood proteins that play a role in regulating the physiological activity of the respective genes.

For question 2
Candidate genes and their molecular signal paths ought to be researched adequately; effect of gene manipulation on certain blood parameters must be known; identification of suitable blood parameters still requires intensive research – parameters must only be modified by doping exclusively and not, for example, significantly affected by physiological factors, such as training, endogenous hormones, foodstuffs, etc.; determination of standard values and defining of upper limits.

Deviations from standard values = perhaps really determinative for "talent."

In-depth text 2: direct detection of gene doping
In direct detection, the doping substance looked for is immediately detected in the sample (blood, urine). In gene therapeutic modification, transgene DNA (tDNA) is delivered to the body in a process called transduction. Structural differences between tDNA and endogenous genomic DNA (gDNA) in principle can be utilized for direct gene doping detection.

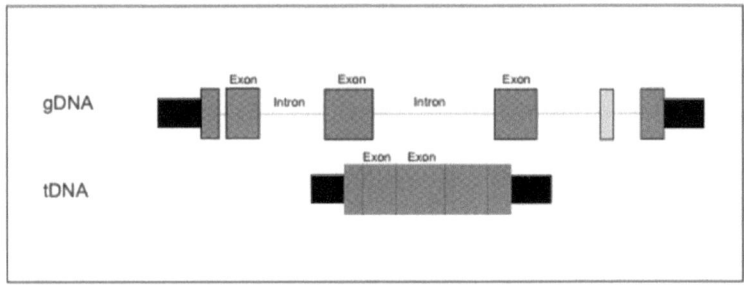

Fig. 1: Exon-intron structure of genomic DNA (gDNA) and transgenic DNA (tDNA). Source: modified from Beiter et al. 2008 Exerc. Immunol. Rev. 2008; 14:73–85

The blueprint for synthesizing proteins is stored in the genes of our gDNA. Besides the coding DNA sequence, these genes also contain large non-coding sequences. The coding gene sequences that carry the information for the amino acid sequence of the expressed protein are called exons. They are interspersed with non-coding DNA sequences called introns. Since, for technical reasons, the length of DNA that can be delivered to the human body is limited, normally gene sequences consisting only of exons are used in gene transfer, since these information units suffice for protein synthesis. Conversely, tDNA in most cases does not incorporate introns. (Fig. 6)

Subsequent to in vivo gene transfer, after a certain time interval the viruses loaded with tDNA, transduced cells and/or free tDNA molecule enter the blood stream. When isolating the entire DNA from a blood sample, it is possible to detect traces of tDNA with a special technique. With help of the polymerase chain reaction (PCR), a molecular genetic laboratory method, defined DNA sequences can be copied with high specificity and made visible.

This permits conclusively distinguishing the minutest amounts of tDNA from gDNA and target them for detection. Since tDNA does not occur naturally in the body, detecting it allows an unambiguous conclusion that a somatic gene transfer has taken place. Therefore, it is feasible to develop a direct detection method for all candidate genes implicated in performance enhancement, one that can be deployed after taking a blood sample as part of a standardized doping control to detect gene doping.

Assignments
1. Describe the differences between tDNA and gDNA and explain how they can be used in direct gene doping detection.
2. Explain the basic principle of direct gene doping detection.
3. Form an opinion on the pros and cons of direct gene doping detection.

Evaluation rubric

For question 1

tDNA is shorter, lacks introns, and is not naturally present in the human body. Because of the absence of introns, established sequence and length differences can be used to differentiate tDNA from gDNA during visualization.

For question 2

Detection of tDNA from blood samples. Blood collection, DNA isolation, copying of tDNA (PCR), visualization of the copied DNA.

For question 3

Pro: conclusive detection of doping -> tDNA not naturally present in the body, therefore no need to check standard reference frames (clear yes / no signal). Detection of the exact doping substance (gene) ≠ generic blood doping or similar, high sensitivity.

Con: Only detectable during a certain period of time (tDNA remains in blood for a limited time only).

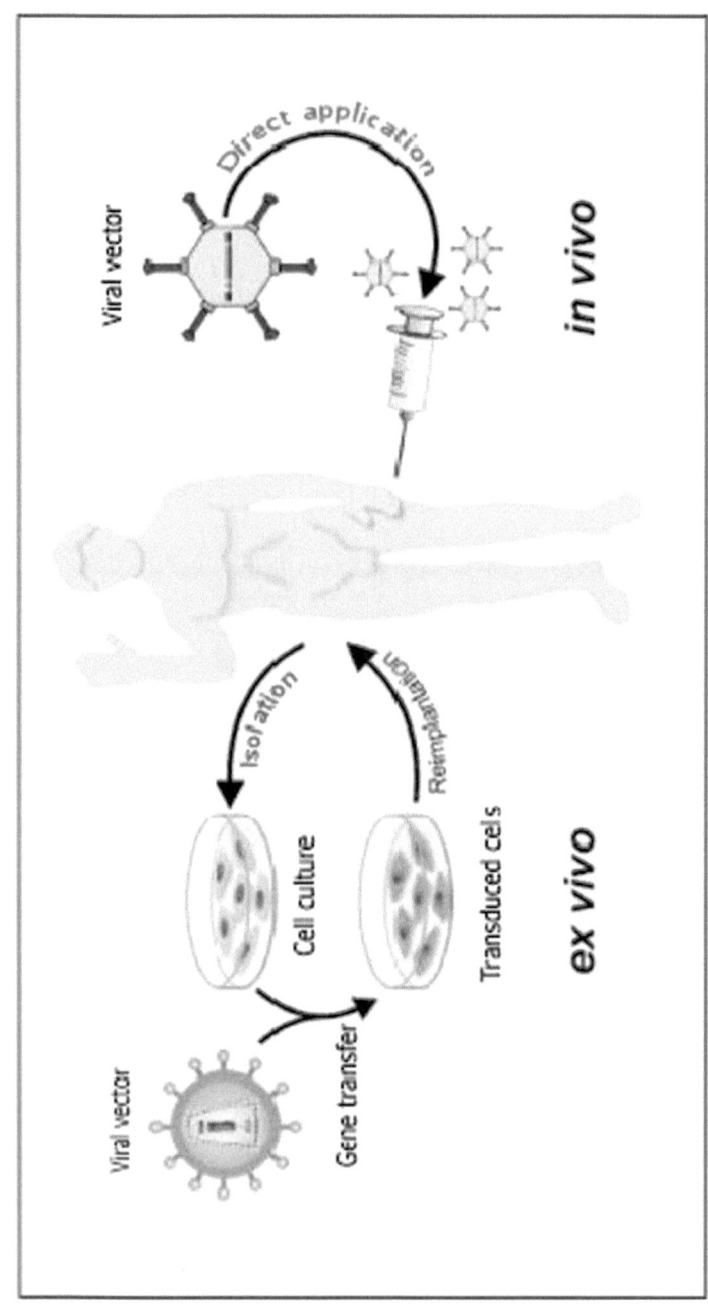

Fig. 2: Options for gene transfers with somatic gene therapy
Source: Beiter & Velders (2012) DZS. Jg 63, Nr. 5. p. 123.

Fig. 3: Cartoon 1
Source: Golombek; 2013

Fig. 4: Cartoon 2
Source: Golombek; 2013

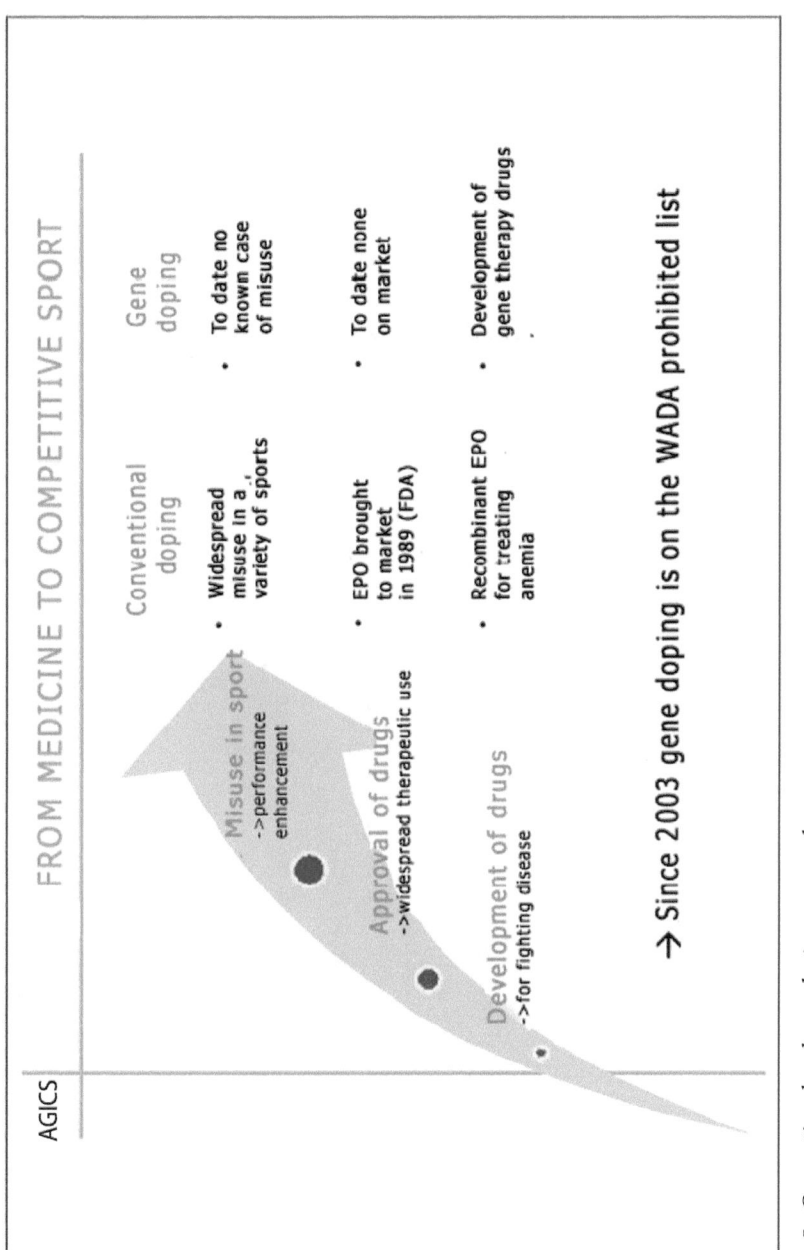

Fig. 5: *Conventional and gene doping compared*
Source: *the authors*

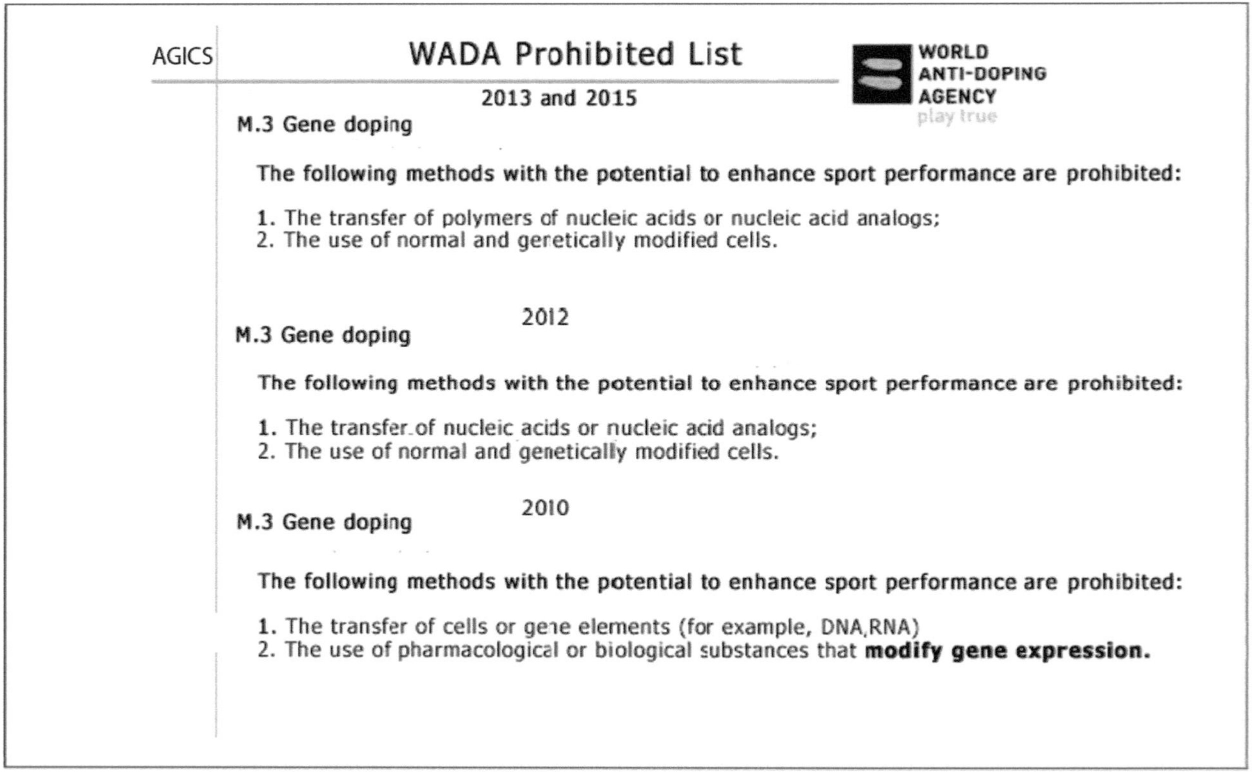
AGICS
WADA Prohibited List
WORLD ANTI-DOPING AGENCY
play true
2013 and 2015
M.3 Gene doping
The following methods with the potential to enhance sport performance are prohibited:
1. The transfer of polymers of nucleic acids or nucleic acid analogs;
2. The use of normal and geretically modified cells.
2012
M.3 Gene doping
The following methods with the potential to enhance sport performance are prohibited:
1. The transfer of nucleic acids or nucleic acid analogs;
2. The use of normal and genetically modified cells.
2010
M.3 Gene doping
The following methods with the potential to enhance sport performance are prohibited:
1. The transfer of cells or gene elements (for example, DNA,RNA)
2. The use of pharmacological or biological substances that modify gene expression.

Fig. 6: Gene doping definition

Source: the authors

7.4 Material for Phase 4

Lead-in to the crisis talk using the scenario (handout)

"Good evening, ladies and gentlemen.

Dateline Frankfurt. Angola's Ery Topoi has won the heavily contested Frankfurt Marathon in one hour 59 minutes 38 seconds and broke the world record by four minutes.

Dateline Tokyo. The favorite in the over 105 weight class from Belarus Myo Statin at the men's weightlifting Asian open championships set a new world record with 270kg in the clean and jerk.

Dateline Oberhof. At the Biathlon World Cup the first day ended after some exciting preliminaries. Attracting a lot of attention – naturally and as many expected – was Alexander Natural. Just a few weeks ago, the top German athlete found himself accused of gene doping due to elevated blood oxygen levels. Natural initially was banned but then was able to prove that the excess red blood cells his body produces naturally are responsible for his elevated oxygen levels. Today was the first time he was qualified to start again. So far during the heats he has not quite come out on top; newcomer Marco Epowitz managed to beat Alexander Natural to the finish line by a whisker. This exceptional athlete has…hold on one second, please…Ladies and gentlemen, I'm just getting the news that there's been an incident at Oberhof…

…that Epowitz has been caught gene doping and has been banned from the competition.

It appears that he used drugs to induce his body to produce more red blood cells as a way of manipulating his oxygen balance.

And [reaching for his ear]…you'll find out more about what is known to date and more background on this surprise development in the crisis talk the World Cup organizers will hold in short order – Subject: *Is it all Epo now? Is it all over for true sport?* We'll bring it to you live after this. So stay with us!"

Evaluation rubric: Lawyer

Legal questions (with answers) for the teacher and/or moderator

The questions below are designed to ensure minimum coverage of the topic's legal aspects and should be asked by the moderator. Decisions on whether and when the questions fit the content and the moment should be made depending on the situation. The row numbers indicate the places in both legal basic texts where the answers to the questions will be found. In addition, the expected answer is given directly below each question. The students have prepared themselves specifically for the questions below. A recommended way to start things off is with questions one and two, saving questions three and four for later in the talk (also refer to: sketch of how the discussion proceeds).

Is Marco Epowitz going to jail? (Text material row: 30–37; 51–57; 101–128)

Yes, if he has a non-trivial amount of doping substances in his possession he faces up to three years imprisonment or a monetary fine. In addition, he faces up to five years in prison or a monetary fine if he defrauded someone; in other words, if he hid his doped-up state from someone who then suffered a financial loss because of it. Apart from that, self-doping is not prosecutable. Our law is designed not to punish self-inflicted harm but harm done to others. Harming others involves someone causing another person injury.

Did Epowitz defraud anyone by having gene doped? (Section 263 StGB) (Text material row: 122–171)

Organizer (+)

Yes, he defrauded the event organizer. His deception resides in not having abided by the rules. The organizer is misled about Epowitz allegedly being "clean" and, because of the deception and error, pays him – provided the doped Epowitz had won – the competition prize money although, based on the competition's rules, he did not earn it. The organizer's property suffered damage by Epowitz being incapable of performing without defaulting on his competition contract.

Sponsor (+)

In addition, Epowitz has defrauded his sponsor. Should the sponsor's contract contain a prohibition against doping, here again Epowitz has lied about being clean of doping. Because the sponsor signed the contract and is not compensated fully for the damage to the image of the convicted Epowitz, he has also suffered a pecuniary loss.

Competitors (–)

Epowitz did not defraud his competitors in the event since they are aware, or, at a minimum, must assume that some of the athletes competing use prohibited means, hence there can be no mistake and a fraud must be excluded. Whether the competitors "know the score" is something difficult to ascertain and to prove. Should they be ignorant of the score, there is another precondition that will negate fraud against the competitors that is inherent in the doped athlete actually receiving his prize money from the organizer's or prize provider's hand. That the second place finisher does not put in his claim is only a supposition, but not – as is necessary for fraud – the flip side of the gain. However, the competitor can file a civil suit to recover the entire amount of the damages that resulted from the worse finishing order.

Spectators (–)

Epowitz did not defraud the spectators. After all, in return for their ticket price they enjoyed an athletic competition, so that they suffered no pecuniary loss.

Can other participants also be subject to prosecution if they gene doped Epowitz? For example, his coach, if he gave him or injected the substance (Text material rows 40–69)

a. *German drug law (AMG)*

Yes, his coach can receive up to a three-year jail sentence or a monetary fine if he injected Epowitz with the substance or gave it to him in the form of pills. This penalty also applies to all pharmaceutical companies that market doping substances, doctors who prescribe doping substances, and any persons that use such a substance on the athlete. Should Epowitz still be a minor, i.e., under the age of 18, then the prison sentence increases from one year to as much as ten years. (Sub-section 6a, 95 AMG)

b. *Bodily harm*

Should the trainer or anyone else damage Epowitz physically (cause bodily harm) by blood doping him, they face up to five years in prison or a monetary fine. The mere attempt is also punishable. However, if Epowitz agreed to the gene doping and is aware of what will be done to him, it makes the perpetrator immune to prosecution. Still, there are limits to consent (agreement).

These limits come into play when doping infringes on public morality, i.e. violates the sense of fairness of all just and reasonable thinkers. Should the gene doping lead to serious health problems, which is definitely likely, then there is a breach of public morality. That makes the perpetrator punishable even when there is consent on the part of Epowitz.

c. *Homicide*

Should someone gene dope Epowitz, and he dies because of it, then the charge becomes homicide. Punishment can range from up to five years in prison or monetary fine for negligent homicide, five years or more for manslaughter, and a sentence of life in prison in case of murder.

d. *Accessory to fraud, bodily harm, homicide and third-party assisted doping*

Anyone who helps with fraud or incites Epowitz to commit fraud also becomes subject to prosecution. Furthermore, anyone is punishable who helps other, third persons in committing bodily harm, homicide or blood doping, either by providing help (aiding) or inciting (abetting) third persons.

What sanctions should Marco Epowitz expect on the level of sports law? What do NADA and WADA provide for in such cases? (Text material rows 180–203)

Gene doping violates the WADA or NADA codes and is prohibited. Depending on the extent of his guilt, Epowitz must be prepared for a temporary suspension, a ban, having his competition results disqualified, the withholding of financial support or the imposition of financial sanctions.

Form your own judgement on the topic of gene doping. What does the legal perspective contribute to it?

Evaluation rubric: Spectator/Fan

For c.) Make a list of pro and con arguments for doping/gene doping

Pro	Con
This is how you please the crowd.	Gene doping destroys sport.
With gene doping, athletes are only obeying the logic of competitive sport	
This is how you boost performance.	With gene doping, it is no longer about the body's performance capability but about the human body's doping tolerance.
Athletes exist to overcome their own limits. Gene doping is nothing unnatural. Gene modification can be viewed as compatible with nature, unlike earlier forms of performance enhancement.	Gene doping is not natural.
Athletes are not natural beings. The question is which technologies enhance athletic performance.	Fatalities, serious illnesses can occur. Gene doping is a misuse of gene therapy. Unforeseeable consequences of gene transfer.
Everything evolves. If sports want to remain relevant, it has to accept the genetically modified athlete.	Is it still sport when engaging in gene doping? Where is the competition? Should the athlete or the molecular biologist get the gold medal?
Everyone dopes, it has been going on forever.	
Lombardi principles: winning is everything. Victory only through gene doping.	Coubertin principle: (Competing is important. Athlete's self-perfection.)
Gene doping brings money and fame.	Fairness and participating are also great values.
If only the few practice gene doping it leads to more unfairness than if the many do it. It would be more honest. This serves to establish a level playing field among athletes. There is no such thing as clean sport.	
No one is interested in sport if records are not broken. There is no ethical consciousness on the podium.	

Evaluation rubric: Athlete/Marco Epowitz

Select the appropriate heading for each of the text snippets lined up one below the other.

Heading 1:   Level playing field through gene doping; talent alone is not enough; doping as the norm.
Heading 2:   Gene doping is something natural – the athlete is not.
Heading 3:   Detection is difficult.
Heading 4:   The sports system inexorably pulls gene doping along with it.
Heading 5:   A great many athletes dope.
Heading 6:   New records achieved through gene doping.
Heading 7:   Glory, kudos, and money seduce into gene doping.
Heading 8:   A great many athletes dope.
Heading 9:   New records only through gene doping.
Heading 10: Glory, kudos, and money seduce into gene doping.

Lines of legitimization/argumentation, justification levels

- Gauge which of the justifications of your actions might be best for you to bring into the discussion and take relevant notes, if need be, on the "Discussion Help" worksheet.

Health

- Athletes relativize health risks. They say that the health-related risks were not known in any case. Even science has not been able to confirm them unequivocally.

Command of one's own body

- Athletes say that they are free to command their own bodies (Constitutional principle).

Accessibility/supply channels

- Athletes point to the ease of acquisition, for example, via the Internet.

Difficult detection

- Athletes point out the difficulty of detecting gene doping. This seduces athletes into misuse.

Ignorance of legal ramification
- Athletes emphasize that there are is as yet no clear basis in law relative to gene doping.
- Athletes say that they were not aware of culpability.

System logic of sports
- The inner logic of the system of competitive sports forces athletes into the gene doping trap. It leaves them no choice but to artificially enhance their performance.

Environment
- The environment (coaches, managers, doctors, etc.) pushes the athletes into rule-breaking behavior.

Financial and psychic costs (also an argument for "environment")
- Compounding of gene doping "preparations" does not cost athletes much. Steep investments in athletes require a high rate of return.
- High (non-financial) investments by athletes should really pay off for them. Gene doping is designed to ensure the likelihood of success.

Level the playing field
- Athletes say that doping helps make up for genetic differences and disadvantages.

Everyone does it, gene doping included
- Athletes perceive gene doping as the norm in their sport.

The goldman dilemma
- Athletes willingly accept that their health may suffer (up to and including death) if it leads to athletic success (gold medal).
- World-class performances (records) are so tightly bunched that the athletes say only gene doping lets them produce stand-out performances.

Evaluation rubric: coaches

(See Spectator/Fan)

Evaluation rubric: federation official

1. List the values and aims of the sport federation

Sport

- Only sport free of doping is credible sport and will survive; doping is nothing more than fraud: zero tolerance for doping.
- Performance, competition, fair play, discipline, team spirit, nonviolence.
- Learn to deal with failure; test your limits; play by the rules; be part of the team; "school of democracy."

Competitive sport

Competition principle; performance principle as basis of civil society; level playing field; learning difficult movement patterns; character building; idea of fair play; practice moderation; positive elite concept; identification with country – patriotism without nationalism.

2. Name the constraints under which each athlete must operate

- WADA (and NADA): The World Anti-doping Agency sets reporting system and catalog of penalties
- Wraparound monitoring with electronic reporting system; elaborate control program (treated like felons?)
- Question if individual rights are violated by the monitoring (e.g., prepare to be surprise tested 365 days a year)
- Athletes compare monitoring to wearing "electronic ankle bracelet"
- The EU data protection law potentially contradicts the code. No final rule has been issued.

(Only distribution and dealing are punishable under an appendix to the German Drug Law.)

Chapter 8
Additional material slide presentation: legal basics of gene doping

The following presentation of the legal basics of gene doping can either be utilized by the teacher as a synopsis of the two legal base texts or by the teacher or student as presentation material.

Compact slide overview

Slide 1
Legal basics of gene doping

Slide 2
The legal picture

We need the law to protect against the misuse of (gene) doping.
 Our Constitution does not proscribe doing harm to self, only doing harm to others (through doping).

Slide 3
The doping prohibition

- **§ 6a AMG:** "Trafficking, prescribing, or using drugs on others for the purpose of doping in sport is prohibited."
- **Potential offenders:** Drug companies, physicians, coaches or other persons

Slide 4
The doping prohibition, ctd.

- Beyond that, the law prohibits possession of doping substances in other than non-negligible amounts.
- **Potential offenders:** All who possess them, hence also the athletes!
- **Sanctions:** Anyone violating the prohibition against third-party assisted doping and doping substance possession is subject to three years imprisonment or a monetary fine (§§ 6a, 95 AMG). When the victim is a minor, the prison sentence increases to ten years.

Slide 5
Criminal code

- **Bodily harm:** § 223 Criminal Code (StGB)
- **Homicide:** § 212 Criminal Code (StGB)
- **Fraud:** § 263 StGB, if a pecuniary loss is suffered by the deceived competition organizer or sponsor

However, the athlete's consent → leads to the perpetrator's impunity in case of bodily harm, but only up to the threshold of public morality, which is crossed when serious bodily harm results.

Slide 6
Sport federation rules

- **WADA Code:** prohibition on (gene) doping
- **NADA Code:** prohibition on (gene) doping

Sanctions:
- Temporary suspension of the athlete
- Permanent ban
- Disqualification of meet results
- Financial sanctions, etc.

Slide 7
How does the law define gene doping?

- The transfer of nucleic acid polymers or nucleic acid analogs
- The use of normal or genetically modified cells

(WADA Prohibited List 2015)

Legal basics of gene doping

The legal picture

We need the law to protect against the misuse of (gene) doping.

Our Constitution does not proscribe doing harm to self, but doing harm to others (through doping).

The doping prohibition

- **§ 6a AMG**: "Trafficking, prescribing, or using drugs on others for the purpose of doping in sport is prohibited."

- **Potential offenders:**
  – Drug companies
  – Physicians
  – Trainers or other persons

The doping prohibition, ctd.

- Beyond that, the law prohibits possession of doping substances in non-negligible amounts.

- **Potential offenders**: All who possess them, hence also the athlete!

- **Sanctions**: Anyone violating the prohibition against blood doping and possession of doping substances is subject to three years imprisonment or a monetary fine (§ 6a, 95 AMG). When the victim is a minor, the prison sentence increases to ten years.

Criminal Code

- **Bodily harm:** § 223 StGB
- **Homicide:** § 212 StGB
- **Fraud:** § 263 StGB, if a pecuniary loss is suffered by the deceived competition organizer or sponsor

However, the athlete's consent -> leads to the perpetrator's impunity in case of bodily harm, but only up to the threshold of public morality, which is crossed when serious bodily harm results.

(Sport) Federation Rules

- **WADA code:** Prohibition against (gene) doping
- **NADA code:** Prohibition against (gene) doping

Sanctions:
- temporary suspension of the athlete
- ban from competition
- disqualification of competition results
- financial sanctions, etc.

How does the law define gene doping?

- The transfer of nucleic acid polymers or nucleic acid analogs

- The use of normal or genetically modified cells.

(WADA Prohibited List 2015)

Student materials
Volume 2

Volume 2 (Student Materials) contains all worksheets required for the teaching unit. They can therefore be printed out together for the comprehensive classic option. Alternatively, individual material can be selected depending on personal preference.

Chapter 1
Scientific basics of gene doping

MATERIAL 1: Basic text
Definition of gene doping

The World Anti-Doping Agency (WADA) defines gene doping as the transfer of nucleic acids or nucleic acid sequences (DNA, RNA) as well as normal or genetically modified cells with the potential to enhance sport performance.

Nucleic acids

DNA is double-stranded and is packed into chromosomes in the nucleus of every cell in our bodies. RNA is produced in the nucleus as a single-stranded copy of DNA; it exits the nucleus and is used in the cytoplasm as pattern for the synthesis of proteins (gene expression). While a living organism's cells contain mostly identical DNA, different cells produce differing RNAs and proteins depending on their function (skin, muscle, liver cells, etc.) (Differential gene expression).

Mutations

 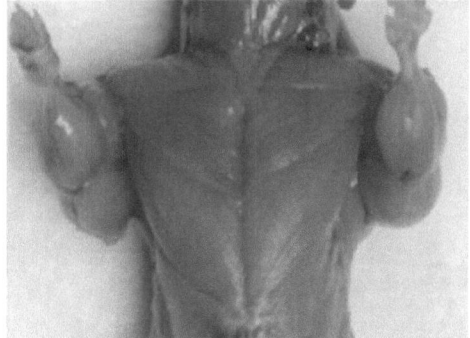

Fig. 1: Genetically modified mouse.
Source: Lee 2007; PLoS One. 29:2(8):e789

Fig. 2: Genetically modified mouse.
Source: Lee 2007, PLoS One. 29;2(8):e789

Mutations in most cases are naturally occurring random changes in the DNA. From mutations it is possible to tell how large the effects can be when a living organism is modified genetically. For example, in Belgian blue cattle, the gene for myostatin, a hormone that inhibits muscle growth, is defective due to a mutation. This causes the animals to grow bigger muscles. In experiments on mice, this effect has already been successfully produced through genetic engineering (Figs. 3, 4).

Gene doping vs. conventional doping

The concept of gene doping is based on the principles of gene therapy. In contrast to conventional doping, gene doping is designed to get the athlete's body to the point where it produces the doping substance (e.g., EPO) itself. Under certain circumstances, after the gene transfer, the body may produce the doping substance for the rest of its life, and, in the worst case, this happens in an uncontrolled manner and leads to severe side effects. There are to date few longitudinal studies on this topic, and the side effects naturally are also highly dependent on the kind of infiltrated gene and its regulation.

Germline therapy vs. somatic gene therapy

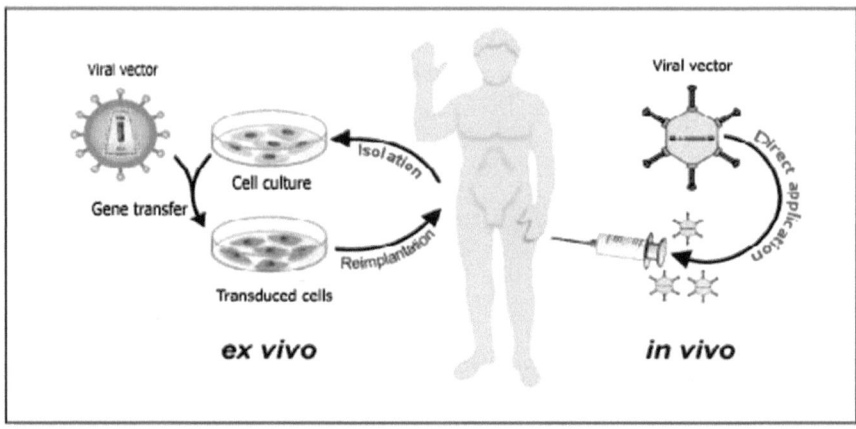

Fig. 3: Options for gene transfers with somatic gene therapy
Source: Beiter & Velders (2012) DZS. Jg. 63, Nr. 5. p. 123

Gene therapy differentiates between **germline** and **somatic therapy**. In germline therapy, germline cells (egg or sperm) are modified genetically. This type of gene therapy can be inherited by the offspring. In somatic gene therapy, (from "soma," the Greek word for body) as the name implies, the body's non-reproductive cells are manipulated genetically – as a rule, this form of gene therapy is not inherited by the next generation. In somatic therapy, the gene transfer with the therapeutic gene (also called a transgene, e.g., EPO) takes place under controlled conditions outside the body (**ex vivo**; see Fig. 3). This type of gene transfer has the advantage that only cells taken from the body are manipulated genetically. In contrast, with **in vivo** gene therapy, the gene transfer occurs inside the body, for example, by injection into the bloodstream or the musculature. The in vivo form of gene therapy may be easier to implement technically than ex vivo therapy, but the side effects of in vivo therapy can be much more severe as the result of a massive immune reaction against parts of the viral vector and, in the worst cases, it can lead to death through multiple organ failure.

In order to get the therapeutic gene into a patient's body, gene therapy makes use of the ability of viruses to penetrate the cell membrane efficiently and in some cases even into the cell nucleus. In this respect, viruses serve as "gene taxis," or in technical language "viral vectors."

Which body cells are particularly suited for gene transfer in gene doping?

This of course depends on which aspect of physical performance capability is to be enhanced. It is possible to increase muscle strength by genetically manipulating muscle cells. However, an in vivo injection, perhaps in the thigh, will only suffice to reach a small part of the superficial muscle fibers. After embryonic development stops, muscle cells are no longer capable of dividing. This offers the advantage that genes infiltrated by gene therapeutic means into the cells remain there if they do not integrate themselves in the chromosomes (episomal maintenance).

However, this property of the musculature represents an inhibiting factor for gene therapy, for example, when viruses are employed that will only integrate with actively dividing cells. Still, in the musculature there are stem cells (satellite cells) that do have the ability to divide and foster muscle growth with their DNA to regenerate muscle after hard training. Only future research projects can show if it is possible to modify these cells genetically in order to boost performance. Since muscles make up about 30–40% of body mass, this tissue is also suited for production of non-muscular gene products, such as EPO. In this case, using an appropriate virus, the EPO gene will be infiltrated into the musculature where it will produce the EPO protein. The EPO will enter the bloodstream as a growth factor to stimulate production of red blood cells (erythrocytes) that will then improve the blood's oxygen transport capacity and hence endurance sport performance.

Transferability of animal experimental findings to humans

Gene therapy has already achieved remarkable results in animal experiments. Transferring them to humans, however, so far only succeeded in a few isolated instances. In animal experiment it has even been shown that transgene activity can be turned on or off as desired by administering drugs (on/off system). Some of the gene transfer studies of humans found unexpected, occasionally grave side effects, such as the emergence of blood cancers (leukemia) or multiple organ failure stemming from severe immune reactions. Before gene therapy can be applied safely more intensive research and development will be needed.

Following successful gene therapy, the body will produce the transgene's gene product (protein) and thereby make it impossible for conventional doping detection methods that are based on detecting artificially produced doping substances to detect the protein. For this reason, gene doping detection must deploy a new method. Various moleculobiological techniques offer themselves in this regard.

Questions

1. Describe the differences between gene doping and conventional doping.
2. Explain possible methods for transferring genes used in gene therapy.
3. Weigh the potential opportunities and risks of gene doping. As you do so, think carefully about what yardsticks to apply for your evaluation.
4. What are the special properties of muscle cells and how do they affect treatment by gene therapy?

MATERIAL 2: in-depth text 1
Direct detection of gene doping

With indirect detection methods, it is not the doping substance to be identified itself that is detected by the test (unlike direct methods) but rather its effects on certain blood values. So, for example, when doping with testosterone is suspected, it is the ratio of testosterone and epitestosterone in the body that is analyzed. Another example is indirect detection of EPO. This is usually done using the hematocrit (volume percentage of red blood cells in blood), the hemoglobin concentration, the number of reticulocytes (immature erythrocytes), macrocytes, and a few other parameters. The standard values differ from individual to individual, even if in medical practice certain average values are specified as being normal.

This can result in a hematocrit value that is read as elevated actually being normal for some people. In this connection, it needs to be asked if it is not simply those naturally caused deviations from standard values that in many cases account for athletic talent.

This requires taking into consideration the differences between individual athletes. Individual blood values observed over longer periods are therefore more meaningful than one-off blood tests. To capture such values is the purpose of setting up blood profiles. This means that blood samples are taken from the athlete at regular intervals and under varying circumstances, for instance before and after training camps, and the results are entered into a blood passport. This makes it possible to flag deviations from the individual value reference range that may indicate doping.

However, for the numerous gene doping candidate genes such standard values are not yet known and great deal of study will be required to understand the molecular interrelations and influencing factors on each of the genes.

A precondition for effective doping detection using indirect testing methods is that all candidate genes and their molecular signaling pathways must be known. As part of indirect detection methods, blood proteins will ultimately be identified that play a role in the regulation of each gene's physiological activity. For example, in cases of manipulating the myostatin gene, changes in the concentration of myostatin inhibiting proteins (e.g., follistatin) might be considered as indirect evidence of doping.

Still, it remains a challenge for anti-doping research to even identify suitable blood proteins for doping detection, because these parameters must invariably only be modified by doping and not by physiological factors, such as physical activity. For this, scientists must determine what concentration is normal to begin with. In determining specific maximum limits, they must establish clearly to what extent the protein concentration can be altered naturally, as for example by training, endogenous hormones or certain foods, etc. It follows that intensive research will still be required in the future to gain the ability to detect doping in general and gene doping in particular.

Questions on the scientific in-depth text
1. Name the differences between direct and indirect doping detection methods.
2. Form an opinion of the challenges posed by direct doping detection. In doing so, keep in mind the characteristics that the particular blood parameters must exhibit.

MATERIAL 3: In-depth text 2

Indirect detection of gene doping

In direct detection, the doping substance looked for is immediately detected in the sample (blood, urine). In gene therapeutic modification, transgene DNA (tDNA) is delivered to the body in a process called transduction. Structural differences between tDNA and endogenous genomic DNA (gDNA) in principle can be utilized for direct gene doping detection.

The blueprint for synthesizing proteins is stored in the genes of our gDNA. Besides the coding DNA sequence, these genes also contain large non-coding sequences. The coding gene sequences that carry the information for the amino acid sequence of the expressed protein are called exons. They are interspersed with non-coding DNA sequences called introns. Since, for technical reasons, the length of DNA that can be delivered to the human body is limited, normally gene sequences consisting only of exons are used in gene transfer, since these information units suffice for protein synthesis. Conversely, tDNA in most cases does not incorporate introns. (Fig. 6)

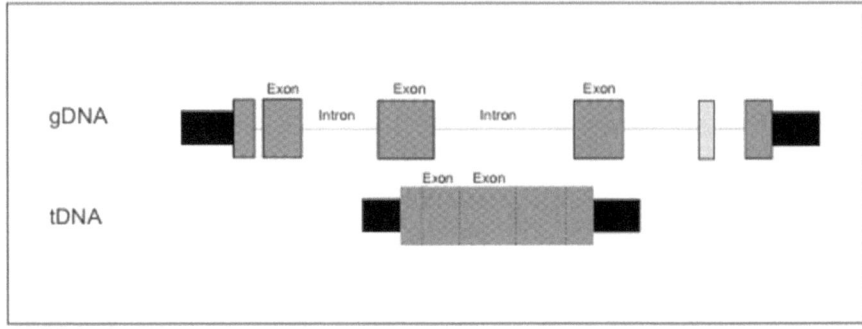

Fig. 1: Exon-intron structure of genomic DNA (gDNA) and transgenic DNA (tDNA). Source: modified from Beiter et al. 2008 Exerc. Immunol. Rev. 2008; 14:73–85

Subsequent to in vivo gene transfer, after a certain time interval the viruses loaded with tDNA, transduced cells and/or free tDNA molecule enter the blood stream. When isolating the entire DNA from a blood sample, it is possible to detect traces of tDNA with a special technique. With help of the polymerase chain reaction (PCR), a molecular genetic laboratory method, defined DNA sequences can be copied with high specificity and to become visible.

This makes it possible to conclusively distinguish the minutest amounts of tDNA from gDNA and target them for detection. Since tDNA does not occur naturally in the body, detecting it allows an unambiguous conclusion that a somatic gene transfer has taken place. Therefore, it is feasible to develop a direct detection method for all candidate genes implicated in performance enhancement, one that can be deployed after taking a blood sample as part of a standardized doping control to detect gene doping.

Questions
1. Describe the differences between tDNA and gDNA and explain how they can be used in direct gene doping detection.
2. Explain the basic principle of direct gene doping detection.
3. Form an opinion on the pros and cons of direct gene doping detection.

Chapter 2
Crisis talk material

MATERIAL 4: News report for the crisis talk

http://www.vidup.de/v/ktKff/

"Good evening, ladies and gentlemen.

Frankfurt. Angola's Ery Topoi has won the heavily contested Frankfurt Marathon in one hour 59 minutes 38 seconds and broke the world record by four minutes.

Tokyo. The favorite in the over 105 weight class from Belarus Myo Statin at the men's weightlifting Asian open championships set a new world record with 270kg in the clean and jerk.

Oberhof. At the Biathlon World Cup the first day ended after some exciting preliminaries. Attracting a lot of attention – naturally and as many expected – was Alexander Natural. Just a few weeks ago, the top German athlete found himself accused of gene doping due to elevated blood oxygen levels. Natural initially was banned but then was able to prove that the excess red blood cells his body produces naturally are responsible for his elevated oxygen levels. Today was the first time he was qualified to start again. So far during the heats he has not quite come out on top; newcomer Marco Epowitz managed to beat Alexander Natural to the finish line by a whisker. This exceptional athlete has …hold on one second, please…Ladies and gentlemen, I'm just getting the news that there's been an incident at Oberhof…

…that Epowitz has been caught gene doping and has been banned from the competition.

It appears that he used drugs to induce his body to produce more red blood cells as a way of manipulating his oxygen balance.

And [reaching for his ear]…you'll find out more about what is known to date and more background on this surprise development in the crisis talk the World Cup organizers will hold in short order – Subject: *Is it all Epo now? Is it all over for true sport?* We'll bring it to you live after this. So stay with us!"

Chapter 3
Infopack for the lawyer role

MATERIAL 5: Introduction to the lawyer's role

Pointers for the role: You are taking part in the experts' round table discussion on the legal background of the gene doping topic. At any time during the discussion the moderator may ask for your input. The questions below will help you prepare for the role. You can find the answers in the two basic texts.

1. **Will Marco Epowitz have to go to jail?** (Text material rows: ____; ____; ____)
   Yes, if _____
   _____ (§§ __ AMG).
   In addition, he faces _____
   _____
   _____ (§ __ StGB).
   Otherwise, self-doping is not punishable. Because: _____
   _____
   _____

2. **Did Epowitz defraud anyone (§ 263, Criminal Code) by having gene doped himself?** (Text material rows: _____)

   a) **Organizer (+)**

b) _____ (+)

c) Competitors (−)

d) _____ (−)

3. **Are others involved subject to criminal prosecution if they gene dope Epowitz? So, for example, his coach if he provided or injected the substance?** (Text material rows: ____; ____) Distinguish between a)–d):

e) **AMG**
   Yes, Epowitz's coach _____

   _____(§§ 6a, 95 AMG).

f) _____

If the coach or anyone else damages Epowitz's health (=_____) by third-party assisted doping then they face

_____
_____
_____
_____
_____

g) Homicide

If anyone gene dopes Epowitz and he dies from it, it is homicide. _____

_____
_____
_____
_____
_____
_____

h) Being an accessory to fraud, causing bodily harm, homicide or third-party assisted doping _____

_____
_____
_____
_____
_____
_____

4. What should Marco Epowitz expect on the level of legal sanctions by the sport? What do NADA and WADA provide for in such cases?
(Text material rows:_____)

_____
_____
_____
_____
_____
_____

5. **Form your own opinion on the topic of gene doping. What does the legal perspective contribute to it?**

_____
_____
_____
_____
_____
_____

MATERIAL 6: Gene doping and the law

Text authored by: Dorothea Magnus, legal expert.

1. Problem

When you play sports, normally you will not have a run-in with the law. On the contrary, laws facilitate playing sports. Every school has physical education classes. That is written in a law – the school ordinance. Many sporting events, competitions and championships are governed by such laws to ensure that they are conducted in a fair manner. It already starts when configuring the events. In other words, there will be certain weight classes for weightlifting, exact distances to be run, swum, or bicycled, and so on. All of it is laid down in law.

Doping is behavior that massively violates the fairness doctrine in sports. Doping endangers the athlete's health, disadvantages the competitors and distorts the athletic competition. It is such serious violation of the law that it is punished not only with monetary penalties but with imprisonment. There are a number of ways to dope, for example, using anabolics, EPO or with brand-new techniques like gene doping. Research into the latter may be in the early stages, but gene doping looks like the doping method of the future. We need the law to prevent misuse of these techniques in a timely manner.

What gene doping is and how the law can protect against it is what we want to talk about with you in this teaching unit.

2. The legal picture

The starting point for our law is that every human being has the right to do as he or she pleases with own self and body. This right does not even end where self-injury begins. This foundation of our Constitution applies to the areas of sport. If people want to damage their own body with doping substances, there is nothing to stop them. The law allows it. What it does not allow is injuring someone else with doping substances. Therefore, our law does not want to punish self-injury but injury caused by third parties. Third party injuries are all those in which someone injures another person.

Just what does the doping prohibition in fact forbid? The prohibition is written into the Drug Law and states:

> *"Trafficking, prescribing, or administering drugs for doping purposes to others is prohibited." Anyone violating this prohibition is subject to imprisonment for up to 3 years or to a monetary penalty (§§ 6a, 95 AMG).*

This prohibition is directed above all at pharmaceutical companies that bring doping substances to market, physicians who use such substances on athletes, i.e., inject them or administer them in the form of pills.

Should the victim be a minor, that is, under the age of 18, then the prison term increases by one to ten years. This takes into account that children especially easily become victims for coaches, parents or other parties that – with the best of intentions – want to enhance their sports performance.

> *Moreover, the law prohibits the possession of doping substances in a non- negligible amount.*

This edict against possession applies to all persons, hence the athletes themselves. It is similar with drugs: Anyone who consumes drugs, for examples, smokes a joint, shoots up heroin, etc. will not be punished. Only dealers, those possessing or storing larger quantities, face prosecution. Possession and doping prohibitions apply to all types of doping, gene doping included.

Should the perpetrator cause harm to an athlete's health (cause bodily harm) through third-party assisted doping, life (homicide), or the property of other parties (fraud), that person will be subject to prosecution. Also subject to punishment under the law is anyone who aids or abets other third persons in doing so. If the athlete acquiesces in the doping aware of what is being done then this basically exculpates the perpetrator from the charge of committing bodily harm. This will exempt from prosecution anyone who third-party doped the athlete (the victim). However, consent does have its limits; namely, when the doping violates common decency, i.e. violates the sense of what is right held by all just and reasonable thinkers. If the doping leads to serious damage to health, then prevailing opinion assumes a violation of public morality. In that case, the perpetrator is subject to prosecution despite the "victim's" consent. Severe, even permanent health impairment is absolutely likely with doping. In the case of gene doping, such damage may even be passed on – through changes in DNA and thus the genetic information – to the doping sinner's progeny.

3. Legal definition of gene doping
How exactly does the law define gene doping?

M3. Gene doping (WADA Prohibited List 2015)
The Word Anti-Doping Agency prohibits the following methods for enhancing athletic performance:

1. The transfer of nucleic acid polymers or nucleic acid analogs
2. The use of normal or genetically modified cells

The most likely areas for implementing gene doping include enhancing the oxygen supply and building up skeletal muscle as well as energy provisioning.

It is still unclear if and to what extent gene doping is being deployed in practice. Methods and technologies also remain in their infancies. However, experts predict more widespread use of a series of cutting-edge substances and methods for targeted influencing of gene activity in the years ahead.

MATERIAL 7: Legal dimensions of gene doping

Text authored by:
Katharina Lammert,
legal expert

Gene doping is the most potent future form of illegal performance enhancement and opens completely new dimensions for fighting it that must be taken seriously. On the legal level, it occurs above all through the application of laws. Laws and the sanctions attached to them, however, can only be applied if proof of gene doping can stand up in court.

In the present case, the detection of gene doping was successful. Marco Epowitz enhanced his performance with the help of this dishonest method. Should he be allowed to continue taking part in competitions? How might his conduct be viewed from the legal standpoint?

The state, which has co-responsibility in the fight against doping, can proceed against gene doping transgressors under the Drug Law (AMG) and the Criminal Law (StGB). In addition to bodily harm and homicide offenses (see Text 1) under the statutory offenses of the Criminal Law the offense of fraud under § 263 StGB could also come into consideration.

(1) Whoever with the intent of procuring for himself or a third person an unlawful pecuniary advantage damages the property of another by pretense of false or misrepresentation or suppression of true facts gives rise to or sustains a falsehood will be punished by up to five years imprisonment or with a monetary fine.
(2) The attempt is punishable (…) (§ 263 StGB – Fraud)

In order to have committed the fraud offense, Marco Epowitz must therefore have deceived someone, causing this person to have a mistaken belief and under its influence made a disposition of property and ultimately suffered a pecuniary loss because of it. In addition, Epowitz must have acted deliberately with the intent of enriching himself.

But whom might Marco Epowitz have defrauded?

With (gene) doping, fraud against the organizer, competitor, fan and the sponsor come into question. The athlete deceives these persons about the fact that he is abiding by the doping regulations and so creates the false impression that he is not doped and is abiding by the doping rules.

The deception of the **organizer** consists of the athlete's declaration that he is playing by the rules. He either gives this assurance explicitly by signing a participation agreement containing such an anti-doping clause ahead of the competition or it is implicit in the doped athlete's participating in the competition. For in doing so, the athlete by his actions declares conclusively that he also accepts the rules, including the prohibition on doping. The organizer now thinks that everything is in good order but is misled about the athlete's alleged "cleanness." Based on the deception and the mistaken belief the organizer will pay the competition prize money to the athlete – provided the doped athlete has won – even though it is not payable under the rules of competition. The organizer's property is damaged in that the doped athlete cannot perform as agreed, i.e., without doping, under the contract concluded with the organizer and the pecuniary loss cannot be compensated by a fitting quid pro quo, i.e., being undoped. Along with that, the athlete must have acted deliberately with intent to achieve an illegal pecuniary advantage. Because of the violation of the rules embodied in the start contract there is no claim to the pecuniary advantage. Hence, fraud against and at the expense of the organizer can be substantiated.

On the other hand, when it comes to fraud against a "clean" **competitor** it is already problematic if there even is a mistake on the part of that competitor. The athletes know the scene and know better than all others who are part of it that not all competitions are contested with allowed means. Competitors must therefore take into account that at least a few of the athletes stepping up use prohibited means, which excludes any mistake and with it any fraud.

Fraud against and to the harm of the **spectators** also falls apart; in this instance, because there is no evidence of a pecuniary loss. This is because the spectators have received a quid pro quo consisting of the completion of the athletic competition for their entrance fee and hence suffered no loss.

When it comes to fraud against a **sponsor**, a deceptive action can be said to occur if there is a doping prohibition written into the sponsorship contract but the athlete nevertheless engages in doping. The sponsor who relies on the contract's contents as a result is mistaken about the athlete's being "clean." The property disposition is seen in the conclusion of the contract by which the sponsor obligates himself to payments to the athlete. The pecuniary loss ultimately results from the circumstance that the sponsor receives no further economically equivalent quid pro quo. For because of the doping, the athlete's image suffers immensely. This could accordingly furnish the grounds for fraud against the sponsor.

If the conditions for a fraud are met, under § 263 StGB the doped athlete is subject to a monetary penalty or imprisonment for up to five years.

How would Marco Epowitz be treated on the sport rules level?

The sport and the state share the important duty of combating doping. Sport has autonomously created the world or national anti-doping codes as the rules framework for the fight against doping. They prohibit self-use of doping substances, evasive actions with respect to their control, as well as possession, trafficking, use on others and complicity in various other acts.

All of these defined breaches of anti-doping regulations cover gene doping, dating back to 2003 when it was already added to the list of prohibited substances and methods (the so-called prohibited list).

> "The following methods with the potential of enhancing athletic performance are prohibited:
> 1. The transfer of nucleic acid polymers or nucleic acid analogs
> 2. The use of normal or genetically modified cells." (Prohibited methods: M3. Gene doping)

Athletes are responsible for all the substances in their bodies. Finding who is at fault for the presence of a prohibited substances is not germane. The precepts contained in the WADA and NADA codes are to be translated by the sports federations into their own rule books. The prohibited list must be anchored in the respective federation statutes. Sports organizations that adhere to the WADA or NADA codes have defined doping inclusive of gene doping as illegal conduct and prohibited it.

For every established violation, a sanction fitting the subjective guilt must be imposed. Among the sanctions enumerated by Article 10 of the NADA code are temporary suspension of the athlete, a ban, disqualification of competition results, holding back financial subsidies as well as the levying of financial sanctions.

Fraudulent acts during sports competition in the form of (gene) doping can also violate both applicable statutory strictures and sanction standards of the sports federations or of superordinate sports institutions like WADA and NADA. Both types of proceedings can be carried out and closed out with sanctions.

Chapter 4
Infopack for the fan/spectator role

MATERIAL 8: Introduction to the fan role

In the discussion on the central question: "Is it all EPO or what? – Is sport finished after the Epowitz incident?" you are going to take on the role of a representative of the public, a spectator. As such, you are to form an opinion about the events swirling around Marco Epowitz in particular and gene doping in general and then speak up for these worked-out positions in the discussion. Of course, you can also raise more questions that matter to you during the discussion.

Prepare in a way so that everyone in the group can join in the "crisis talk."

Read MATERIAL 9–11 on the gene doping topic and complete the following assignments to prepare for the discussion:

a. Draw up a list of pro and con arguments for doping/gene doping
b. Agree among each other on **one** position as "the public" whether you want to accept gene doping or not.

The following questions should also help to pin down a position:

1. Give reasons for why you might still watch competitions on TV if (gene) doping were legal.
2. Substantiate whether documentaries titled "Gene doping: attack of the mutants" could lead to social marginalization and/or stigmatization of convicted athletes.
3. Opine whether there should be special performance categories for "gene doped" athletes? What does the public expect from competitive sport?
4. Give reasons for why you might think of gene doping as a natural outgrowth of technology already in use in competitive sports. Do you regard gene doping as morally reprehensible? If so, why?

MATERIAL 9: The F.A.Z.[6] in conversation with Prof. K.-H. Bette

Interviewee:
Karl-Heinrich Bette, Sports sociologist, Technical University Darmstadt

FAZ: *Who encourages athletes to dope instead of pointing out the possibilities of coming out on top without doping?*

Bette: For sociology, doping is an interesting subject because by this example we can show how strongly individual action is influenced by social conditions. The first factor that affects the athlete is the logic of competitive sport itself. The Olympic motto is "citius, altus, fortius," i.e., "faster, higher, stronger." Looking at this motto, we become aware of its inner infinitude which is stored symbolically. You are never satisfied with what you have attained. The competitive sports system is oriented toward escalation. To keep upping performance is what in the end fascinates the public.

FAZ: *Who are other significant actors?*

Bette: In competitive sport, the public can vicariously experience more intensity of life, variety, community, celebration and exhilaration; it can worship heroes and participate in a world of uncertainty but one that remains without consequences for the individual. Because of the public interest, other actors pay attention, foremost the mass media that want to boost circulation or audience share. Given the media presence and through the media's presentation of athletic performance, political and business entities are interested for the sake of increasing their own electability or using sport for advertising purposes.

FAZ: *In other words: the public is part of the doping problem?*

Bette: Absolutely. The important thing is that the public's entanglement in the doping problematic operates primarily on an unconscious level. When a million spectators turn on the TV, a dynamic is set in motion that is completely outside the will of the individual fan. Even doping opponents in this way can become part of the problem. Beyond this, the discussion of doping among the general public is conducted in a rather ambivalent manner. Many could care less, others are horrified and disillusioned. The public is torn between its own need for diversion on the one hand and sticking with the traditional rules of sport on the other.

---

6 Frankfurter Allgemeine Zeitung (Frankfurt's daily newspaper).

FAZ: *Would lifting the prohibition on doping solve the problem?*

Bette: No way! Liberalization would only make the problem worse. All we would have left in competitive sports are the reckless risk takers. We also would no longer just test the athletes' performance capability, but only the tolerance limits of the human body to doping. Many stakeholders would turn away in disgust. My guess is acceptance of doping would completely destroy sport.

(FAZ, 03.07.2008. Available on the <u>FAZ website</u> – *accessed on 1.17. 2013*)

MATERIAL 10: The following are excerpts from talks with two well-known experts on the subject of "gene doping in sport"

Taking part in the conversation:
Andy Miah, bioethicist, Theodore Friedmann, molecular biologist

Miah: I say that sport needs gene manipulation. The question is not if we should use it in sport, because athletes are there to surpass human limits. It's what competitive sport is all about. We almost impose performance enhancing technologies on athletes to make them break through their limits.

Many people assert that to modify genes is to play God. If that's true, we've been doing it for a long time. I simply view it as a new way of working with technologies to make us and our living conditions better. Biology is advancing and how we influence it is an important factor in this development. That's not something unnatural. Indeed, gene modification can be viewed as compatible with nature, something that earlier forms of performance enhancement were not.

Friedman: *I think it's absurd when some members of the pro gene doping society say that it's simply a natural continuation of technology and that genetically engineered athletes are natural. I can't see anything natural in that whatsoever. The pressure in sport and fact that there a circles that push for gene technology in sport I find to be a scary, dangerous combination.*

We know from gene therapy studies that some very grave things have happened to patients. Including some fatalities. The technology can be used for healing illnesses. But if the intent becomes changing a natural predisposition in a human being, let's say a young healthy athlete and to seriously damage him in doing so, that should not be allowed.

Miah: Athletes already live in a technological culture. For the past two and half thousand years. They use technologies ever since the Olympic Games in Greece. The standing long jump advanced because the athletes held weights in their hands to give themselves more momentum. Modern athletes today are simply part of a sophisticated technological evolution. We must quickly discard the notion that athletes are natural beings, and then we've got to talk about which technologies enrich sports performance. I think those are the technologies that let athletes train harder and achieve better performances.

Friedman: *The contention that athletes early on should be pulled into the development of genetic methods is completely wrong. It completely contradicts what we know about the difficulties created by trying to heal a child with a weak immune system. This clinical work is very difficult. To put athletes in this position and expose them to the risks and imponderables of gene transfer is unethical.*

Miah: I see the discussion about performance enhancement in competitive sport in a larger context. We already live in a performance enhancement culture. And society is going to utilize this technology in all kinds of areas. The question is, will sport we exempted from that, or will it continue to be relevant and adapt. When gene technology finds ways to make non-therapeutic implementation safer, then sport also will have to develop and accept the genetically engineered athlete.

Friedmann: *If gene therapy is to be used in sport, we need to ask ourselves if that is still sport. Where is the competition? Who wins the gold medal? Is it the athlete or the molecular biologist sitting on the sidelines?*

(Arte, 2010. Gene doping: Attack of the mutants. Arte website - *accessed on 01.17.2013*).

MATERIAL 11: "Unrestricted doping is fairer"

Interviewee:
Franz Begov, Sports historian

FOCUS Online: *Herr Begov, among German cycling pros one doping admissions follows another. The cycle pro Jörg Jaksche even calls the doping system fair because "everyone dopes." Can competitive sports still be saved?*

Begov: Look, they doped in classical antiquity already. One of the most successful athletes of the Greek Olympic scene, the wrestler Milon of Croton ate great quantities of bull's testicles and bean-sized capon stones, so-called alectoria from young, castrated roosters. So he took substances comparable to anabolic steroids.

Ingesting such animal products was not proscribed. Also, the victor in the 1904 Olympic Marathon was full of strychnine and alcohol without being disqualified.

FOCUS Online: *History here or there – is that a fitting example for explaining the misuse of performance enhancing drugs in modern sport?*

Begov: To put is simply, there are two concepts of sport that fundamentally differ from each other. On the one hand rules the Coubertin principle. According to Pierre de Coubertin, the founder of the International Olympic Committee (IOC), the uppermost action maxim of the athlete is not in the winning but in competing – and the athlete's self-perfection, physically and morally, that is tied to it. But, in fact, in sport you come across another personality type much more frequently: the athlete who acts according to the Lombardi principles. Lombardi was an American football coach who declared winning to be the most important if not the only goal in sports competition.

FOCUS Online: *Why is the Coubertin type in the minority in professional sport?*

Begov: For the pros, sport is always also about making money and making a life. Yardsticks other than fairness or taking part are more important. Winning brings with it with numerous benefits: besides monetary and material ones, there's also the chance of social advancement. So that favors the type who has signed on for the Lombardi model. And that makes him do a lot for a win – even if that means doping. The result: when two or more Lombardi-type athletes encounter each other, doping eventually is unavoidable. Because the winning paradigm dominates, not the precept of fairness. This is why there are cases of doping all the time.

FOCUS Online: *So for the athlete it's all about winning – but how about for the spectator?*

Begov: Think about what happened with the Americas Cup in sailing. With the German team bobbing around in next to last place, i.e., without success, the TV transmissions cut away. Why? Because it interests no one when someone brings up the rear. That includes the media – and the viewers no less. They want the spectacular, success, the victory event. In my opinion, victory in many types of sports is only possible with prohibited drugs and painkillers: whether with health damaging attacks on the opponent, or with illegal drugs or painkillers. This is why I think gene doping will never be brought under control. The athletes and their handlers are always way ahead of the control methods. Besides: even if it were possible to catch a large part of the dopers, there would always be a few Lombardi types left over. Their victory prospects would consequently improve considerably. And, in principle, this will lead to more unfairness than if everyone doped.

FOCUS Online: *So, then the logical conclusion in your view can only be that all athletes should dope.*

Begov: Even if it were the better solution from the standpoint of sports ethics that no one dopes: if you subscribe to the Lombardi model – winning at all costs –a level competitive situation will only come about if everyone is allowed to dope. That way, at least, you've created something like a rudimentary balance among athletes. I take that to be the more honest solution. Because only then the best can prevail again – and not the one best able to circumvent the control measures. After all, the athletes still have to turn in a formidable performance.

FOCUS Online: *It does not worry you that it could be linked to fatalities?*

Begov: Sure it does. Because that's risk that would come along with lifting the restrictions. But I simply cannot believe that "clean sport" would be feasible in the future – or that the Coubertin type will prevail in the long run. No matter where you look in competitive sports: it's the quest for success using all means available. Already a fourth place finish no longer counts.

FOCUS Online: *But a sport in which there is open doping would turn off the fans.*

Begov: I doubt that there is much of a pronounced ethical consciousness in the stands. The public enjoys the event, the happening per se – and above all the action.

FOCUS Online: *What you're saying there sounds like an athletic declaration of bankruptcy.*

Begov: You can think that way if you follow the pure sports line of thinking. But when you link sport above all with money, media presence and the like, other yardsticks apply. Look at the sums that winners in world-class sports earn. Look at the disregard of a cultural heritage as with the Olympic stadium in Munich. They made it into a single-purpose soccer pitch. That documents impressively which laws shape sport. Ethical appeals won't get you far here.

FOCUS Online: *It sounds like doping is a thoroughly human trait.*

Begov: Let's put it this way: it is a human trait to want to keep outdoing yourself and some people will go farther than others in that regard.

(Focus Online, 07.18.2007, http://www.focus.de - *accessed on 01.17.2013*).

Chapter 5
Infopack for the athlete role

MATERIAL 12: Introduction to the athlete role

In the discussion on the central question: "Is it all EPO or what? – Is sport finished after the Epowitz incident?" you are going to take on the role of the athlete (Marco Epowitz) who was caught gene doping. Your PR advisor has pulled together a text in a briefing package to help you justify your actions in the impending "crisis talk." Because it was set up on short notice, your PR adviser was unable to think of any pithy headings to help cut right to the heart of the text compilation.

Prepare in a way so that everyone in the group can join in the "crisis talk."

Read the materials and complete the following assignments in preparation for the talk, using the discussion structuring help sheet.

1. Think up suitable headings for each of the text excerpts lined up one below the other(Example: Rubric No. 1 = Level playing field through gene doping.).
2. For more intense preparation for the discussion also work on the following questions/assignments:
   a. Decide which of the justifications of your actions that are quoted in in the material you can best bring into the discussion and, for example, take appropriate notes on the AB "Discussion Helper."
   b. Pressure from which source (officials, coaches, media, and competitors) had the most/least influence on you?
   c. What are your hopes/fears with regard to your conduct?

MATERIAL 13: Athlete

Heading 1:

- The best training does little to help make the leap into the world elite if nature doesn't take a hand in it. The best athletes have a special aptitude, they have winner muscles." (Spiegel Online, 2008 - *accessed on 01.17.2013*).
- "Maybe somewhere out there is a mega-talent that can make it to the top without doping. I don't want to exclude that, but the odds I'd say are very low. In many kinds of sports you hardly have a chance without doping to even get close to the world's best. Let's call it like it is." (Spiegel Online, 2008 - *accessed on 01.17.2013*).

- "The way I see it, doping in competitive sports is a form of support. This form of effort is per se unhealthy and even a minimal support can help make the sport somewhat more bearable. I'm aware that this can only kick off the discussion. Both extremes – lifting all restrictions for one and clean sport for another – are unrealistic. "Doping light" under medically controlled conditions could be a possible way. But the question always crops up of the role model effect for the kids. We are stuck in a dilemma here and I don't have an answer for what the way out is." (ibid.)
- "It may be perverse, but the doping system is fair, because they all dope. Cycling sport without doping is only fair if no one really dopes any longer. A cyclist told me that there are supposed to be deals between a few teams and the World Cycling Federation. There you have to assume that there won't be a general change of thinking. The cyclist was proud telling me this. That's when I knew: nothing has changed." (Spiegel Online, 2007 - *accessed on 01.17.2013*).
- "I had it up to here, getting beaten time and again, and I didn't get what made the guys from the USA run so fast. (…) Your talent will only get you so far. Today I know what else you can do. (Stern, 2003 - *accessed on 01.17.2013*).

Heading 2:
- "Basically, it's impossible to estimate the health risks from misuse relying on clinical drug trials." (TAB, 2008 - *accessed on 01.17.2013*).

Heading 3:
- "Many people assert that to modify genes is to play God. If that's true, we've been doing it for a long time. I simply view it as a new way of working with technologies to make us and our living conditions better. Biology is advancing and how we influence is an important factor in this development. That's not something unnatural. Indeed, gene modification can be viewed as compatible with nature, something that earlier forms of performance enhancement were not." (Miah, A. Arte, 2010 - *accessed on 01.17.2013*).
- "Athletes already live in a technological culture. For the past two and half thousand years. They use technologies ever since the Olympic Games in Greece. The standing long jump advanced because the athletes held weights in their hands to give themselves more momentum. Modern athletes today are simply part of a sophisticated technological evolution. We must quickly discard the notion that athletes are natural beings, and then we've got to talk about which technology enrich sports performance. I think those are technologies that let athletes train

harder and achieve better performances." (Miah, A. Arte, 2010 - *accessed on 01.17.2013*).

Heading 4:
- The starting point for our law is that every human being has the right to do as he or she pleases with own self and body. This right does not even end where self-injury begins. This foundation of our Constitution applies to the areas of sport. If people want to damage their own bodies with doping substances, there is nothing to stop them. The law allows it. What it does not allow is to injure someone else with doping substances Therefore, our law does not want to punish self-injury but injury by third parties. Third party injuries are all those in which someone injures another person. (Magnus, 2012, unpublished manuscript)
- Just what does the doping prohibition in fact forbid? The prohibition is written into the Drug Law and states: "Trafficking, prescribing, or administering drugs for doping purposes to others is prohibited." Anyone violating this prohibition is subject to imprisonment for up to 3 years or to a monetary penalty (§§ 6a, 95 AMG). (Magnus, 2012, unpublished manuscript)
- "Since these repressive measure in the fight against gene doping will be very elaborate and are still connected to a series of open questions of law, it is unlikely that they will suffice to deter gene doping. Concepts of gene doping prevention must be brought in." (TAB, 2008)
- "Within the framework of publicly subsidized sport, abiding by the WADA and NADA rule book is being demanded in the meantime from the payment recipients." (ibid.)
- "Sports organizations that have incorporated the WADA or the specific NADA code for Germany into their internal organizational rule book or have made analogous arrangements have formally prohibited their members from gene doping. This applies to large parts of competitive sport but not individual sports carried on in fitness studios." (ibid.)

Heading 5:
- "Detection of gene doping probably will be even more costly than detection of current doping practices." (ibid.)
- "However, specific risks can be derived from the principles of methods for targeted modification of gene activity, yet without an empirical basis they merely represent scientifically plausible assumptions." (ibid)

- "The detection of non-viral vectors ("naked" DNA, siRNA) could be even more difficult in view of the short half-life values of nucleic acids. It is completely unclear what detection would look like in procedures in which cells are taken from the body, are genetically engineered outside the body and are then returned to the body." (ibid.)
- "By the time of the Olympics, experts reckon with the first athletes armed with gene doping – seeing that myostatin blockers at best are detectable by muscle fiber biopsy, but by contrast not in blood or urine." (Focus Online, 2008 -*accessed on 1.17.2013*).
- Zabel: "I doped because it works." (Spiegel Online, 2007 - *accessed on 1.17.2013*).

Heading 6:
- "In view of the order of magnitude that doping has reached in sport, pointing to the deviant behavior of individual athletes is not enough. For a comprehensive understanding of doping behavior social contexts must be addressed. These include, for example, the global commercialization of performance and competitive sport: sport itself has become a business and for many athletes it has become their profession. The media and the expectations of a global audience created the preconditions for this and they intensify the economization process of athletic performance. That makes 'winning at any cost' all the more important. The dominance of the performance imperative connected with the prospect of winnings condition structures that are open to anything that will help boost performance." (TAB, 2008 - *accessed on 01.17.2013*).
- "In the sports system, the sports organizations are the actors that seek to mediate between the performance and success demands of the athlete's surroundings – politics, media, sponsors, public – and the athlete herself: They foster the motivation and capability of their athletes and they organize competitions for performance comparison. Their position and their influence in the overall scheme of things therefore depend on the success of their athletes. To that extent they are also – like the athletes – stuck in a kind of 'doping trap.' They have to ensure a 'clean,' by-the-book performance sport by taking an active role in the fight against doping. But with controls and sanctions they also tend to imperil their success. Much of what is done and not done by the federations with respect to doping is best explained by this being enmeshed in the 'system logic' of competitive sport." (ibid.).
- "Doping is as old as competition itself. There's always someone who wants to win a bit more than the next guy. It's that way not only in sport. You have the same thing in our society. Voluntary motivation is often not on display. In my

opinion, athletes who have been training for years want to get to the top. They often know that it's not possible without help." (T-Online, 2011 - *accessed on 01.17.2013*).
- Jörg Jaksche: "Cycling really is not that much fun. It always hurts. With the sport comes a lot of pain, physical pain. Training is the attempt to enhance your performance capability, so you don't fall behind, and first, so it wouldn't hurt, there was cortisone, then Epo, and today it's fresh blood. Cycling is tough sport. As a soccer player, you can run around the pitch like a dope for 90 minutes, then you score the decisive goal in overtime and you're a hero. In cycling, you get left in the dust in 99 out of 100 races, even when you give it everything. It hurts the entire time, and still you seldom see success." (Spiegel Online, 2007 - *accessed on 01.17.2013*).
- "Exaggerated is relative. Once you've been captured by the competitive sports system, doping is part of it just like breakfast. It becomes the "new normal" and it is just that, normal. Behind the façade, in this house that outsiders will never be able to enter, there is open talk about doping. You exchange know-how." (T-Online, 2011 - *accessed on 01.17.2013*).
- "The doping scene is a part of competitive sports that you cannot look at in isolation. Competitive sport is really a gathering of the insane who move within a sick system. Still, we have to keep in mind that these crazies in their system week after week put on a show that millions follow on their screens or in the stadiums." (ibid.)

Heading 7:
- "How eager the handlers of the top athletes are for new, undetectable gene doping substances was documented in the case of Thomas Springstein. The German track and field coach who made the New Brandenburg sprinter Katrin Krabbe ('it doesn't just come from the spinach') fit and two years ago was slammed with a 16 month jail sentence for giving doping substances to minors, is said in emails to even have gone after the blood doping drug Repoxygen. This is stuff that switches on the Epo production on the gene that was only cleared for use on mice – and about which no one knew how to shut it off." (Focus Online, 2008 - *accessed on 01.17.2013*).
- "After their doping confession, the Festina cyclists accused the team's management. Brochard explained yesterday that management had introduced doping. 'In reality, I feel myself to be a victim of the first order,' the cycling pro wrote in a statement made public by his attorney. He was going to be a co-plaintiff in the case." „ (Welt Online, 1998 - *accessed on 01.17.2013*).

- "Herr Spilker from time to time would invite me to lunch. During the conversations, doping came up more and more often, but at first Herr Spilker did not offer me any doping substances outright. As time went by, he would say, if someone wanted it, he could get a hold of some. In other words, he only painted a picture of the possibility. At first, I also was not interested, I really didn't want to. The talk kept coming back to doping substances. Then came a time when I was frustrated with my athletic performance. Sometime during the 1986/1987 indoor season I actually did consent to it. Due to a problem with my foot at the time, I also was behind in my training." (NZZ Online, 2001 - *accessed on 01.17.2013*).

Heading 8:

"In the area of active international competitive sports a current study showed that in some countries more than sixty percent of the track and field athletes engaged in blood doping or doping with EPO microdoses." (Deutsches Ärzteblatt, 2012 - *accessed on 01.17.2013*).

Heading 9:
- "Every athlete dreams about pushing the boundary higher, faster, farther, set a world record; this is motivation, because it proves that you're the best. Doping helps with that." (Spiegel Online, 2008 - *accessed on 01.17.2013*).
- "The recorded best mark is held by the Cuban Javier Sotomayor; it is at 2.45 meters, set on July 27, 1993 in Salamanca, Spain. 15 years ago." (ibid.)
- "Eike Onnen, age 26, tried everything. He lost five kilos, because light weight flies better. He only weighs 85 kilos, at 1.96 meters tall; his body fat percentage is five. He has improved his approach to 7.5 meters per second. Javier Sotomayor was faster: 8.4 meters per second. It's enough to drive you up the wall." (ibid.)
- "Onnen looked at the second piece of paper a long time, then he said: 'I would really like to break the world record, I want to get into the history books, but I need twelve centimeters. I have no idea how to make that work." (ibid.)
- "A world record is an epic mark, the greatest and athlete can achieve, it's worth more than an Olympic gold medal or one at a world championship. Because a world record stands as a symbol for the humanly possible, of the limit to performance. But where does that lie?" (ibid.)
- "Records excite me, they fascinate the public, they bring the media." (ibid.)

Heading 10:

- "The [athletes] already now are moving at the limits of what is possible through training and conventional doping. To go it one better, all that would remain to also manipulate genetically." (GenEthics Network, 2011 - *accessed on 01.17.2013*).
- "To avoid the threatened loss of much money and honor." (Labo.de, 2011 - *accessed on 01.17.2013*).
- "What Florence Griffith-Joyner said, according to Darrel Robinson: 'When my goal is to make millions, I can afford to invest a few thousand dollars in trying to reach that goal.'" (Focus Online, 2007 - *accessed on 01.17.2013*).
- "Studies for a long time have shown that athletes are prepared to sacrifice a few years of their lives if it will guarantee them an Olympic gold medal. The risk acceptance is generally very high; everyone thinks that it "can't be all that bad." Precisely here is where the responsibility of the physicians lies and also of the handlers, like I used to be. About gene doping, I can only say that perhaps that's where the future is. I don't know of anything specific at this time that fits the definition of gene doping. (T-Online, 2011 - *accessed on 01.17.2013*).

MATERIAL 14: Discussion structuring help

	Headings from the briefing package	Explanations
1		
2		
3		
4		
5		
6		
7		
8		
9		
10		

Chapter 6
Infopack for the role of coach

MATERIAL 15: Introduction to the role of coach

In the discussion on the central question: "Is it all EPO or what? – Is sport finished after the Epowitz incident?" you are going to take on the role of coach to Marco Epowitz. As such, you will be linked in the public's eye with the events involving your protégé.

Agree among each other if you want to take the position as a coach who is accepting of gene doping or not. You can then develop this stance in the subsequent "crisis talk."

Prepare in such a way so that everyone in the group can join in the "crisis talk."

Read MATERIAL 16–18 on the gene doping topic and work on the following assignments so that you can take a stand in the discussion:

a. Draw up a list of pro and con arguments for doping/gene doping.
b Agree among each other on a position for the role of coach, i.e., whether you want to accept gene doping or not.

The following questions are designed to help you prepare even more intensively for the discussion:

c. In MATERIAL 19 examples of genetic mutations are cited. Evaluate whether it is justified when human beings try to compensate for a genetic disadvantage through gene doping.
d. Is gene doping an incremental continuation of the technology that even now is being utilized in competitive sport? Do you think gene doping is morally repugnant? Why or why not?
e. How do the modern forms of the hero cult relate to the notion of sport as "being there is everything."

MATERIAL 16: Excerpts from interviews with two well-known experts on "gene doping in sport"

Taking part in the conversation:
Andy Miah, bioethicist, Theodore Friedmann, molecular biologist

Miah: I say that sport needs gene manipulation. The question is not if we should use it in sport, because athletes are there to surpass human limits. It's what competitive sport is all about. We almost impose performance enhancing technologies on athletes to make them break through their limits.

Many people assert that to modify genes is to play God. If that's true, we've been doing it for a long time. I simply view it as a new way of working with technologies to make us and our living conditions better. Biology is advancing and how we influence it is an important factor in this development. That's not something unnatural. Indeed, gene modification can be viewed as compatible with nature, something that earlier forms of performance enhancement were not.

Friedman: *I think it's absurd when some members of the pro gene doping society say that it's simply a natural continuation of technology and that genetically engineered athletes are natural. I can't see anything natural in that whatsoever. The pressure in sport and fact that there a circles that push for gene technology in sport I find to be a scary, dangerous combination.*

We know from gene therapy studies that some very grave things have happened to patients. Including some fatalities. The technology can be used for healing illnesses. But if the intent becomes changing a natural predisposition in a human being, let's say a young healthy athlete and to seriously damage him in doing so, that should not be allowed.

Miah: Athletes already live in a technological culture. For the past two and half thousand years. They use technologies ever since the Olympic Games in Greece. The standing long jump advanced because the athletes held weights in their hands to give themselves more momentum. Modern athletes today are simply part of a sophisticated technological evolution. We must quickly discard the notion that athletes are natural beings, and then we've got to talk about which technologies enrich sports performance. I think those are the technologies that let athletes train harder and achieve better performances.

Friedman: *The contention that athletes early on should be pulled into the development of genetic methods is completely wrong. It completely contradicts what we know about the difficulties created by trying to heal a child with a weak immune* system. *This clinical work is very difficult. To put athletes in this position and expose them to the risks and imponderables of gene transfer is unethical.*

Miah: I see the discussion about performance enhancement in competitive sport in a larger context. We already live in a performance enhancement culture. And society is going to utilize this technology in all kinds of areas. The question is, will sport we exempted from that, or will it continue to be relevant and adapt. When gene technology finds ways to make non-therapeutic implementation safer, then sport also will have to develop and accept the genetically engineered athlete.

Friedmann: *If gene therapy is to be used in sport, we need to ask ourselves if that is still sport. Where is the competition? Who wins the gold medal? Is it the athlete or the molecular biologist sitting on the sidelines?*

(Arte website, 2010. "Gene doping: Attack of the mutants." - *accessed on 1.17.2013*).

MATERIAL 17: Andy Miah on ethical reflections on gene doping

1. Many technologies are developed in the first instance to help sick people. But there are also examples of how the use by healthy people advances the technologies for use with the sick, for example, in plastic surgery. It was commercialization that powered its development. So, I can't see why the intent of helping the ill and suffering should be at odds with the wish by the healthy to profit from them as a "lifestyle application." The two can complement each other.
2. As a society, we, after all, want to continually improve the quality of life – beyond therapeutic measures, even beyond what is natural. We want people to be fit, make it possible for them to engage in pursuits that they enjoy.

(Arte, 2010, Gene doping: attack of the mutants.)

MATERIAL 18: "Unrestricted doping is fairer"

Interviewee: Franz Begov, Sports historian

FOCUS Online: *Herr Begov, among German cycling pros one doping admissions follows another. The cycle pro Jörg Jaksche even calls the doping system fair because "everyone dopes." Can competitive sports still be saved?*

Begov: Look, they doped in classical antiquity already. One of the most successful athletes of the Greek Olympic scene, the wrestler Milon of Croton ate great quantities of bull's testicles and bean-sized capon stones, so-called alectoria from young, castrated roosters. So he took substances comparable to anabolic steroids. Ingesting such animal products was not proscribed. Also, the victor in the 1904 Olympic Marathon was full of strychnine and alcohol without being disqualified.

FOCUS Online: *History here or there – is that a fitting example for explaining the misuse of performance enhancing drugs in modern sport?*

Begov: To put is simply, there are two concepts of sport that fundamentally differ from each other. On the one hand rules the Coubertin principle. According to Pierre de Coubertin, the founder of the International Olympic Committee (IOC), the uppermost action maxim of the athlete is not in the winning but in competing – and the athlete's self-perfection, physically and morally, that is tied to it. But, in fact, in sport you come across another personality type much more frequently: the athlete who acts according to the Lombardi principles. Lombardi was an American football coach who declared winning to be the most important if not the only goal in sports competition.

FOCUS Online: *Why is the Coubertin type in the minority in professional sport?*

Begov: For the pros, sport is always also about making money and making a life. Yardsticks other than fairness or taking part are more important. Winning brings with it with numerous benefits: besides monetary and material ones, there's also the chance of social advancement. So that favors the type who has signed on for the Lombardi model. And that makes him do a lot for a win – even if that means doping. The result: when two or more Lombardi-type athletes encounter each other, doping eventually is unavoidable. Because the winning paradigm dominates, not the precept of fairness. This is why there are cases of doping all the time.

FOCUS Online: *So for the athlete it's all about winning – but how about for the spectator?*

Begov: Think about what happened with the Americas Cup in sailing. With the German team bobbing around in next to last place, i.e., without success, the TV transmissions cut away. Why? Because it interests no one when someone brings up the rear. That includes the media – and the viewers no less. They want the spectacular, success, the victory event. In my opinion, victory in many types of sports is only possible with prohibited drugs and painkillers: whether with health damaging attacks on the opponent, or with illegal drugs or painkillers. This is why I think gene doping will never be brought under control. The athletes and their handlers are always way ahead of the control methods. Besides: even if it were possible to catch a large part of the dopers, there would always be a few Lombardi types left over. Their victory prospects would consequently improve considerably. And, in principle, this will lead to more unfairness than if everyone doped.

FOCUS Online: *So, then the logical conclusion in your view can only be that all athletes should dope.*

Begov: Even if it were the better solution from the standpoint of sports ethics that no one dopes: if you subscribe to the Lombardi model – winning at all costs –a level competitive situation will only come about if everyone is allowed to dope. That way, at least, you've created something like a rudimentary balance among athletes. I take that to be the more honest solution. Because only then the best can prevail again – and not the one best able to circumvent the control measures. After all, the athletes still have to turn in a formidable performance.

FOCUS Online: *It does not worry you that it could be linked to fatalities?*

Begov: Sure it does. Because that's risk that would come along with lifting the restrictions. But I simply cannot believe that "clean sport" would be feasible in the future – or that the Coubertin type will prevail in the long run. No matter where you look in competitive sports: it's the quest for success using all means available. Already a fourth place finish no longer counts.

FOCUS Online: *But a sport in which there is open doping would turn off the fans.*

Begov: I doubt that there is much of a pronounced ethical consciousness in the stands. The public enjoys the event, the happening per se – and above all the action.

FOCUS Online: *What you're saying there sounds like an athletic declaration of bankruptcy.*

Begov: You can think that way if you follow the pure sports line of thinking. But when you link sport above all with money, media presence and the like, other yardsticks apply. Look at the sums that winners in world-class sports earn. Look at the disregard of a cultural heritage as with the Olympic stadium in Munich. They made it into a single-purpose soccer pitch. That documents impressively which laws shape sport. Ethical appeals won't get you far here.

FOCUS Online: *It sounds like doping is a thoroughly human trait.*

Begov: Let's put it this way: it is a human trait to want to keep outdoing yourself. And some people will go farther than others in that regard.

(Focus Online, 07.18.2007 - *accessed on 01.17.2013*).

MATERIAL 19: "Gold in the genes"

Success in the genome: Numerous geneticists by now are studying the biological causes of athletic competitive performance. They want to know what makes American Michael Phelps a swimming sensation. They ask themselves why it is that Africans often are world class runners. And they are hunting for opportunities to give natural talent a leg up through gene doping. For example, the Finnish cross country skier Eero Mäntyranta has victory in his blood. Thanks to a genetic mutation, his body produces more of the EPO blood formation hormone. Erythropoietin doping is a favorite with sprinters, marathon athletes and cyclists because it boost the red blood cell count way above the norm. The body can therefore transport a great deal more oxygen. Nature gave it to him as a present from birth on, helping him win gold twice at the 1964 Winter Olympics. A similar instance is that of a female athlete from Berlin, who, however, wants to remain anonymous. Through a mutation of the myostatin gene her body builds more than the normal amount of muscle. She has a professional sports career. And when she had a son just five years ago, the mutation emerged to an even higher degree in him. "The boy has double the normal muscle mass," reports Markus Schülke, a doctor at Berlin's Charité, who went public with the case at the end of June. (Excerpted from: Robert Thielicke: Gold in the genes, in: Focus, 23.08.2004, p. 68)

Chapter 7
Infopack for the role of federation official

MATERIAL 20: Introduction to the role of federation official

In the discussion on the central question: "Is it all EPO or what? – Is sport finished after the Epowitz incident?" you are going to take on the role of a representative of the German Ski Federation. Your federation advocates for the interests of German ski sport nationally and internationally. As federation representative you are to form an opinion about the events swirling around Marco Epowitz specifically and gene doping in general. Of course, you can also raise more questions that matter to you during the discussion.

Prepare in a way so that everyone in the group can join in the "crisis talk."
The following questions are designed to help you prepare for your role.
Read MATERIALS 21–25 and jot down key points that you can elaborate on during the discussion:

1. Name the values and goals of your sport federation.
2. Name the requirements that each athlete must meet and opine on how these requirements should be viewed from your federation's perspective.
3. To what extent do the pressure to perform and succeed fit and/or contradict the values worked out above? Give reasons.
4. What is the significance to those participating in the sport of the above proclaimed and required goals and values?
5. Ponder what role your federation plays or can play in connection with the "Epowitz" case specifically and gene doping generally.

MATERIAL 21: The German Olympic Sports Confederation (DOSB) and its self-image

Values in sport
They say that sport is a mirror of our society. True enough, but it is only half of the story. Sport is also part of our society. It not only reflects, but shapes and influences the whole. This is why it is important to cultivate and foster values that are intimately tied to sport: performance, competition, fair play, discipline, team spirit and non-violence. Everyone can contribute here: the competitive athlete as a good role model for young people (Horst Köhler, DOSB image brochure. P. 9).

Throughout history, modern competitive sport has managed time and again in most impressive fashion to unpack a special cultural meaning that can point the way for modern civilization. Despite all the dangers that raise their head today especially in competitive sports, it can continue to provide an ideal model for a competitive society. In it are brought together symbolically most clearly the principles of a performance society, the performance and competition principle and the principle of equal opportunity. Of symbolic meaning here is also the fact that what matters in competitive sports is the learning of and drilling in of difficult movement patterns and abilities. Often it takes months-long, sometimes even years of training, for an athlete to master a skill, a technique or a tactical variation, before he can become a highly skilled master at it. Sports moves must be "achieved" in the truest sense of the word, and such achieved moves become means for expressing the human personality. The performance principle in competitive sports in this regard is especially important for our civil society because inherent in it is the central criterion for the way individual opportunities and available gratifications are determined solely and exclusively by the performance principle. Not favoritism, age, birth, gender or inherited privileges, chance, faith, origin or skin color should determine the individual's position in our society. In an enlightened civil-democratic society there can be no alternative to the performance principle, which, of course, requires social safeguards and an ethical grounding.

Required for nurturing this principle are symbolic forms of passing on and intermediation. Competitive sports in this regard offer a particularly suitable example. If this view of the performance principle and competitive sports is apt, then it is to be welcomed that individuals express this principle symbolically through sport. Competitive sport is therefore worth promoting and it conforms to a necessary societal interest. For this reason, promotion of competitive sport ought also to be striven for going forward, however, without denying the problems and dangers that come with competitive sports.

Our society relies on individual motivation and willingness to perform. Both traits must be promoted emphatically. These characteristics can be expressed in competitive sport, but they can also be discredited in it. This is very frequently the case today. Questions of sports ethics for this reason will materially influence the future development of competitive sports. They are a challenge for athletes, trainers, coaches, and officials. The idea of competitive sport is fundamentally connected to the idea of fair competition. The integrity of competition that is embodied in the self-regulation of competitive sport requires special protection. Only then can the idea of fair play unfold its teaching value. For educational and ethical reasons, we need competitive sport that has learned reasonable moderation and that does not give up its principles in favor of a fascination with record performances. For that, we need officials, trainers, coaches, sports teachers and athletes who feel bound by a sports ethos and make their influence felt in order to make possible in competitive sports those experiences that could be of significance to the development of our society (Helmut Digel, in: DOSB image brochure, p. 45 ff).

MATERIAL 22: The state's goals for competitive sports

The DOSB has made it its mission to assert Germany's position in world sport. All efforts focus on a successful outcome at the Olympic Games. To that end, the "DOSB new competitive sport control model" was adopted at the DOSB member convention in December 2006. The DOSB exercises its control functions in competitive sports through a multi-disciplinary strategic influencing of the Germany-wide competitive sport system. With this control model the top federations are to be equipped to optimally design competitive sports in their area of responsibility. At the 2006 Winter Games in Turin, the German team took first place; in the 2008 Beijing Olympics it won 16 gold medals and was an overall fifth in the unofficial national standings by total medal count for having won a further ten silver and 15 bronze medals. For the London Olympics in 2013, the German Olympics team plans to once again lock up one of the top five nation spots – totally in keeping with the Olympic performance ideal of "citius, altius, fortius." (DOSB image brochure, p. 49)

MATERIAL 23: Excerpts from the DOSB "State Goals for Sport position paper"

Value communication
Sport gives young people especially the chance to test their limits. Tied into this is learning how to cope with failure, respect for opponents, playing by the rules, and acting as part of team. Sports clubs are "schools of democracy" since they offer many opportunities for participating – especially for youth. In a large measure, they convey the values of our society.

Representation and identification
Sport is a lived commitment to performance and personal responsibility. Competitive sport has role model effects with respect to the performance thought and communicates a positive elite concept. Successful male and female athletes inspire, motivate, and can contribute to a positive picture of Germany through their representation function. As an example, let the 2006 soccer World Cup be

mentioned here. Sport offers the citizen identification with their country and its symbols – patriotism without nationalism.

The doping debate

Only sport free of doping is a credible and viable sport. Doping is nothing but fraud – fraud on the competitor and on self. "We declare war on manipulation in any form, be it doping or corruption, with "zero tolerance," declared DOSB President Thomas Bach in his speech to the DOSB founding convention on May 20, 2006.

In the preamble to its constitution, the DOSB accepts the international anti-doping regulations, specifically the World Anti-Doping Code and dedicates itself to humane sport that is free of manipulation and doping. In December 2006, the DOSB member convention adopted the anti-doping "Ten Point Action Plan for Sport and State" by an overwhelming majority. It envisions increasing the control density as well as reinforcing prevention.

Doping prevention is a decisive factor in the successful educational work against doping. The DOSB delegated leadership for it to the German Sport Youth (DSJ) in 2007 and banks on the latter's competence in preventive measures (DOSB image brochure, p. 37).

MATERIAL 24: Information about athlete whereabouts and accessibility

Athletes that have been nominated by their international sport federation or their national anti-doping organization as candidates for a pool of athletes to serve as training controls by sharing precise and up-to-date details about their location and accessibility. The international sport federations and national anti-doping organization coordinate the naming of the athletes and capture of data on location and accessibility and transmit these to WADA. In turn, WADA allows other anti-doping organizations that, under Article 15, are authorized to carry out anti-doping controls on the athlete access to the data. These data are at all times treated as confidential; they are used exclusively for planning, coordination, and implementation of doping controls and are destroyed as soon as they are no longer needed for this purpose. (World Anti-Doping Code [Version as of January 1, 2004], § 14.3)

MATERIAL 25: Op-ed on doping in the F.A.Z.

Written by:
Evi Simeoni, journalist, Frankfurter Allgemeine Zeitung

The competitive athlete made of glass

The greater the threat, the more massive the counter measure: that is exactly how it is with the doping problem in competitive sports. But all-round monitoring of athletes is one means that could develop such considerable side effects that it might lead the fight against doping into a dead end.

Competitive athletes by now have to submit to elaborate controls
How serious an illness is, the layman sometimes can tell only by the medicine that is deployed against it: the greater the threat, the more massive the antidote. That is exactly how it is with the doping problem in competitive sports.

Just how alarming its dimensions are is best realized by the countermeasures taken by the sport itself. The means – comprehensive monitoring of athletes by utilizing an electronic reporting system – however could develop such serious side effects that it leads the doping fight into a dead end. Elite athletes by now have to submit to such an elaborate control program that some among them are starting to rebel against it, even though it raises suspicion that they are trying to sabotage the self-cleaning program of their game.

Several legal fronts
The complaints that athletes are being treated like felons, that they are being pursued by a kind of clandestine service, are not just of the verbal sort. By now, the courts are occupying themselves with the code of the World Anti-doping Agency, which, among others, sets the reporting system and catalog of penalties. EU specialists, too, are taking a closer look at the set of rules with international validity claim that came into effect on January 1 in a revised version. The doping regulators have to defend themselves on several legal fronts: The issue is whether the system violates the individual rights of athletes. Even the EU work law could nullify the regulations since 365 day readiness for surprise doping tests makes no provision for a vacation. But the biggest danger for the code looms from the side of the European privacy protection law. It could tip as soon as the bureaucratic mills finish grinding – and set the doping fight back dramatically.

General suspicion is by now part of the self-image
It has come to this point: Competitive sport – a globe circling, lucrative show business – is fighting strictly on the defensive. The public has lost all confidence

in it and it must stabilize its ethical-moral and educational claim, which assures it of societal acceptance, through a permanent intense control activity. In doing so, it has likely lost sight of the commandment of proportionality. The general suspicion against which aficionados, defenders and profiteers of competitive sports reflexively put up a fight by now is part of the self-image. This is not remarkable: After all, athletes in many disciplines – even where no money is to be made – have proved that they are ready to exploit every opening to doping fraud even with such undignified methods as attaching urine bags to their own body or even exchanging of this fluid through a catheter. Observers as much as athletes determined to put on a genuine performance have had to learn that there is apparently always somebody ready to transgress the limits of rules and self-respect.

Staged cops and robbers game
As a consequence, the top players of the game have to submit contractually to an electronic monitoring system that resembles a punitive measure for law breakers on parole. Their national anti-doping agency becomes Big Brother who constantly wants to know where they can be found and around the clock for 365 days a year at that. Even more, those representing particularly threatened sports, if they are to avoid what in the past often was frequently a staged cops and robbers game, have to peg an hour every day of the year at which the doping trackers can find them without fail at a specified location. Athletes tired of documenting their day-to-day life, yielding up their own basic rights, have even asked to be fitted with an electronic ankle bracelet – that is how matter of course the rigorous control measure have by now become for them.

Societal withdrawal of affection
It would be a bitter point in the history of sport, when of all things the wild determination to wage war against doping would bring home to competitive sports its impotence in the face of this problem. That you cannot ask sport to be an island of the blissful is something its functionaries have affirmed for decades already. But now it sits there like an island of the self-fixated caught in their own bad habits.

    The German lawmaker – accompanied by the intensive lobbying work of the German Olympic Sports Confederation – has not even seen fit to assign to the doping problem the importance that would be required for it to be made part of the criminal code. Just passing on and trafficking are subject to punishment under a section of the drug law. Society's ostracism of pharmacological sport fraud therefore is materially stronger than the legislature's. But competitive sport has to fight the societal withdrawal of affection with all its might, because lose it and you lose your livelihood. (F.A.Z., 2009 - *accessed on 01.18.2013*)

Chapter 8
Supplemental material

MATERIAL 26: Audience assignments

1. Think what question you as an audience member could ask of one or more discussants during the crisis talk:

   _____
   _____
   _____
   _____
   _____

2. Take notes on how the subject of gene doping is viewed by the different roles (athlete, coach, spectator, etc.) during the crisis talk (write on the back if necessary).

   _____
   _____
   _____
   _____
   _____

3. Give the reasons if you think Marco Epowitz should go to jail.

   _____
   _____
   _____
   _____
   _____

4. What sanctions should Marco Epowitz be prepared to face on the sport legal level? What do NADA and WADA provide for in such cases?

   _____
   _____
   _____
   _____
   _____

MATERIAL 27: Your bottom line

1. You have worked intensively on the subject of gene doping. If you were now going to post a comment on the subject on the Internet (anonymously or perhaps on your Facebook page) on which side of the issue would you come down?

_____
_____
_____
_____
_____
_____
_____
_____
_____
_____

2. Formulate two questions dealing with gene doping that were of most interest to you or would interest you going forward.

_____
_____
_____
_____
_____
_____
_____
_____
_____
_____

MATERIAL 28: Quiz

Question 1:
Which physiological area is **not** a pertinent starting point for possible gene doping?

a) Skeletal muscle
b) Energy provisioning
c) Thermoregulation
d) Oxygen supply

Question 2:
When was "gene doping" added to the WADA Prohibited List?

a) January 1st, 2003
b) June 1st, 2007
c) January 1st, 1998
d) July 1st, 2004

Question 3:
What is a part of the gene doping definition in the most current version of the WADA Prohibited List?

a) use under a doctor's supervision
b) the transfer of the smallest possible particles
c) the transfer of nucleic acids
d) the use of non-organic elements

Question 4:
What kinds of detection methods exist?

a) Behavioral observation and questionnaires
b) Screening systems and indirect detection systems
c) Particle and density measurements
d) Screening systems and direct detection systems

Question 5:
From the perspective of sports law, a principal problem with gene doping resides in

a) legally defensible detectability
b) differentiating victim and perpetrator
c) the extent of the effect
d) the doctor's confession

Question 6:
Gene doping

a) is a pervasive phenomenon in competitive sports
b) has most future potential for enhanced competitive sports performance
c) can be gauged exactly
d) carries no demonstrable health risks

Question 7:
In the ethical debate over gene doping in competitive sport, those opposing gene doping rely above all on arguments

a) are meaningless to athletes and coaches
b) of public opinion
c) by sports officials
d) for protecting life and preserving the sports ethos

Question 8:
In competitive sports, performance enhancement is

a) only important to athletes and coaches
b) incidental
c) the highest internal norm
d) simple

Question 9:
What is **not** a basis for institutional or state prosecution of gene doping violations?

a) the World Anti-doping Code
b) the German Drug Law
c) the German Social Code
d) German Criminal Code

Question 10:
Conventional doping and gene doping are most aptly differentiated by

a) the duration of the effect
b) its applicability to men
c) animal research
d) the lack of fairness

Question 11:
"What is natural?", "How do you differentiate decisions and actions as either better or worse?" are questions that are (primarily) addressed by which discipline?

a) Law
b) Ethics
c) Sports science
d) Medicine

Question 12:
Media reporting of the gene doping topic

a) always delivers a precise recap of the scientific facts
b) points to problem solutions
c) has no influence on forming public opinion
d) tends to simplified and/or exaggerated depictions

Question 13:
The potentially greatest dangers of gene doping are

a) incalculable long term consequences
b) short term effects
c) not known at this time
d) high financial costs

Question 14:
Participation by an athlete with an advantageous but natural gene defect

a) requires free gene doping to create a level playing field
b) has never happened
c) does not violate the provisions of the WADA Code
d) violates the provisions of the WADA Code Prohibited List

Answer 1: c)

The pertinent starting points for possible gene doping are those areas that materially limit or influence human performance capability: skeletal muscle, energy provisioning, and oxygen supply.

Answer 2: a)

Gene doping was added to the WADA Prohibited List on January 1, 2003 as a prohibited doping method.

Answer 3: c)

In the most current version of the WADA prohibited list, under "gene doping" the following methods are prohibited: 1. The transfer of polymers of nucleic acids or nucleic acid analogs; 2. The use of normal or genetically modified cells.

Answer 4: d)

There are indirect detections methods, so-called screening systems, and direct detection methods.

Answer 5: a)

Gene doping poses the challenge of the so-called legally defensible detectability for sports institutions, since detectability provides the grounds for any sanctions.

Answer 6: b)

Gene doping is viewed as the form of performance enhancement with the greatest future potential. The development of new technologies is also inherent in the improvement logic of competitive sports. While still in the early stages, the development of gene therapeutic methods for use in humans holds out the prospect of technologies for performance enhancement that could be misused for gene doping.

Answer 7: d)

Condemnation and rejection of gene technological enhancements in the special case of gene doping rests primarily on two argumentative clusters: protection of life and the sports ethos.

Answer 8: c)

Competitive sports represents a distinct part of modern society that elevates improvement to the highest internal norm within a rigid competitive and record-chasing logic ("higher-faster-longer").

Answer 9: c)

Gene doping infractions are sanctioned on the institutional level by the World Anti-Doping Code; on the state level, by legal proceedings based on the Drug Law and the Criminal Code.

Answer 10: a)

Conventional doping and gene doping are most aptly differentiated by the how long their effects last.

Answer 11: b)

Questions as to the "naturalness" (as opposed to "artificiality") as well as to the (moral) judgment of decisions and actions are key problems of ethics.

Answer 12: d)

Reporting by the media on the gene doping subject matter tends to veer toward simplified and/or exaggerated depiction, for instance, by prominently featuring the cloned super athlete figure as a realistic possibility.

Answer 13: a)

The greatest danger that comes with gene doping are the unforeseeable long tem consequences not only for the health of the athlete but even his genetic material.

Answer 14: c)

Athletes with a congenital gene defect already made headlines in the past because they absolutely gave them an edge in competition. Nevertheless, a congenital gene defect, contrary to use of gene doping, does not violate the WADA prohibited list. The problem of the level playing field, however, remains very controversial.

MATERIAL 29: Glossary

Advantage
Advantage (lat. utilis: beneficial) is a basic concept in economics and the utilitarian ethic that refers to benefits, well-being, and good fortune reaped by one more individuals. The concept of advantage can basically be brought to bear in explaining economic decisions (economy) or in weighing moral decisions (utilitarianism). Accordingly, in economics an action is advantageous if it results in the greatest benefit (most often monetary/material). Utilitarianism, by contrast, holds that ethical decisions in choosing particular options are morally defensible as long as they lead to desired objectives being maximized. Transferred to the field of sports, for example, the economic interest of athletes, of coaches, and sponsors, etc. can explain why gene doping is practiced in competition (success in sport as financial advantage/benefit). If, on the other hand, the goal is to stage a sport for maximum appeal to the public, then the use of gene doping in an enhancement logic (higher, faster, farther) can optimize this goal and therefore represent an advantage as well.

Biotechnology
The exploitation of biological processes for industrial and other purposes, especially the genetic manipulation of microorganisms for the production of antibiotics, hormones, etc. and of foodstuffs (e.g. brewing beer).

Competitive sport
Competitive sport is a segment of modern society that has elevated performance improvement to its highest internal norm under a rigid record-seeking and competitive logic and that distinguishes between legitimate and illegitimate (doping) forms of performance enhancement. Top performing athletes in competitive sports are faced with the challenge of following its rules and prohibitions while simultaneously striving to live up to personal, institutional, media, political, and economic performance aspirations. Doping as a special form of deviant behavior (see also "deviancy") has become structurally established in competitive sport because it offers a way of coping with these competing expectation. A future breakout in use of gene doping, assuming the availability of suitable methods and substances, is not inconceivable.

Damage
Damage, a concept originating in law and economics, is defined as any loss to person or property suffered as the result of a specific event. In keeping with this, genetically engineered enhancement may, for one, damage the athlete's own health, set in motion damage to the image of sports/the particular sport, and/or cause losses to the participating actors because of the public's diminishing interest.

Detection methods
The doping control system classifies as detection methods all valid tests or methods that can detect manipulation with prohibited substances and methods (see the WADA prohibited list). Direct detection methods are distinct from indirect ones. Direct detection immediately detects the prohibited substance in the blood or urine sample. This contrasts with indirect methods, which detect manipulations by identifying deviations from the normal physiological state of a wide variety of molecules (e.g., proteins, DNA, RNA). Indirect methods say nothing about the substance or the method used, however. For this reason, indirect detection methods are also called screening or biomonitoring systems. A blood passport for athletes would be an example of a screening system in which all measured blood parameters regarded as relevant are entered, thus letting modifications over longer time frames be determined. Basically, both methods qualify for detecting gene doping. The development of these detection methods is still in its

infancy. For direct detection, for example, it would be necessary to recognize the introduced genetic elements themselves or the vectors used to introduce them. It is theoretically possible but translating it into practice raises a host of difficult issues. Difficulties arise not just on the moleculobiological level of the misused genetic engineering techniques themselves (for example, differentiating vectors) but also with regard to the multitude and complexity of potential starting points for genetically engineered manipulation in the area of gene modulation. For this reason, most research project concentrate on developing indirect methods. One thing that must be taken into account here, too, is clearing up questions about what the relevant measurement units and time intervals would be.

Deviancy
Deviancy (or deviance) denotes the aberration of individual behavior from socially-developed structures of expectation; in other words, it is the visible disregard of social norms, values, and rules. Doping is a sport-specific form of socially deviant behavior. Doping (and hence potential gene doping) disregards the elemental commandments of fairness and level playing field in athletic competition by using methods and substances prohibited in the WADA code in order to enhance performance.

Enhancement
In general, enhancement is defined as anything done to increase the quality of a healthy life, such as improving cognitive abilities (neuroenhancement) or having plastic surgery done. Enhancement in this sense pervades competitive sport with its inherent logic of improvement ("higher, faster, longer"). In the public's perception, as within the sports system, a distinction is made between legitimate enhancement techniques (e.g., diet, training regimen, mental coaching) and illegitimate ones like doping. Where one starts and the other ends is often fluid. The dividing line hence is tentative and requires constant checking. Gene doping can thus be described as a prohibited sub-category of genetically engineered enhancement.

Enhancement principle
Enhancement is one of modern society's key principles. By way of examples, business is oriented toward enhancing profits and science seeks to enhance knowledge to the maximum. In competitive sports, enhancement has two aspects: maximization and optimization. Maximization aims at quantitative forms of enhancement. Optimization connotes an additional subliminal moral dimension. Enhancement in competitive sports therefore is understood as quantitative-maximal and moral-optimal performance. The modern Olympic motto ("higher,

faster, farther") and the records mentality of competitive sports express the enhancement principle. In this context performance enhancement can, for one, be legitimately attempted and attained (e.g. through training management, nutrition, equipment, etc.) or, on the other hand, illegitimately (with doping).

EPO
EPO (Erythropoietin) is a hormone produced naturally in the human kidney and in small amounts by the liver. It consists of a protein and several carbohydrate chains (glycoprotein). When oxygen shortage (hypoxia) sets in, EPO production increases and is distributed in the blood throughout the body. By acting on stem cells in the bone marrow, EPO stimulates red blood cell (erythrocyte) production. An increase in red blood cells enhances the amount of oxygen the blood can carry. This feature makes EPO a potential gene doping candidate gene, particularly for endurance athletes.

Equity
Coming to grips with the classical concept of equity (justness, fairness) is among the main concerns of ethics, the social sciences, and jurisprudence. Because of the concept's multilayered nature, we can distinguish among various kinds of equity, such as political, social, distributive, and compensatory equity. In the modern world, everyone accepts this conception of equity, regardless of its justification, expressed in the shared fundamental norm: All human beings are without exception equally valued, i.e., they are possessed of equal dignity and to be treated accordingly. Depending on the detailed interpretations of the equal dignity norm under the various conceptions of equity (political, etc.), different values of what is appropriate or just in each case result. According to the social liberal concept, for example, the order of things is just when all disadvantages that people incurred through no fault of their own are compensated for as far as is feasible and can be normatively justified. In doing so, it largely expects people themselves to bear the consequences of the choices they make and of their deliberate actions. These same equity criteria lay claim to moral validity in other ways as well (see also "morality"), and so apply to the sports sphere also. In this instance, equity specifically describes equality of opportunity in athletic competition as overcoming of discrimination and compensation of disadvantages. In addition, the creation of rules equally applying to everyone is linked herein to the chance to achieve a position in competition that is commensurate with abilities and efforts. Genetically engineered enhancement achieved by the athlete's deviant behavior (see also "deviancy") interferes with equality (of opportunity). By so doing, the athletically talented (see also "talent") cannot prevail against doped competitors

despite competent training, and so they are deprived of their deserved position. The perspective on the equity concept shifts, however, if genetically engineered enhancement is viewed as compensating for genetically-conditioned disadvantages, thus underscoring the subject's ambivalence.

Fairness
The concept of fairness, also called fair play, derives from the English word "fair," synonymous with "proper," "just," or "decent." Fairness in sports means keeping to the rules of the game and not using prohibited means such as doping substances. But there is more to fairness than just playing by the rules. Fair also means regarding the sports competitor as a partner with the same rights, and, if called for, helping him or her with an equipment defect or an injury and not using it to press your advantage.

Fraud
Under the German Criminal Code, procuring an unlawful pecuniary gain is considered fraud and is made a punishable offense under § 263 of the Code. In cases of fraud in professional sports, financial asset-related aspects can also play a role, as when a winner's purse is obtained by dishonest means (e.g., doping, bribes). Independent of monetary value-related deceptions and means, however, other instances of fraud in sports that are closely tied to the respective rules of a sport also play an important role. Massive infractions of the rules that violate the tension between the competitive sports principles of a level playing field and the drive to win satisfy the elements of an offense of (sports) fraud.

Gene
A functional segment of DNA that codes for an inheritable structure or function that contains the instructions for producing a particular protein or enzyme.

Gene expression
Gene expression consists of all processes that lead to building a DNA-coded gene product. In the case of protein-coding genes, these include the processes for copying DNA to make RNA (transcription), subsequent modifications of RNA (post-transcriptional regulation), and making proteins on the ribosomes (translation). In the case of non-protein-coding genes, translation is skipped since the RNA is the already completed gene product. Different cell types (muscle, skin, liver, etc.) exhibit differing gene expression patterns (number and strength of expressed genes) enabling the cells to perform their varied functions (differential gene expression). In gene therapy, gene expression of the transgene must proceed normally for the particular (trans)gene product to perform its desired function.

Gene technology/genetic engineering
A sub-area of biotechnology that uses methods allowing the genotype (DNA/RNA) to be isolated, analyzed, copied, and newly assembled. Genetic engineering methods are used, for example, in the pharmaceutical industry to manufacture drugs (e.g. insulin). Genetic engineering research is the source for methods that potentially could be misused for gene doping purposes.

Gene therapy
The concept of gene therapy dates back nearly four decades. It pursues a strategy of introducing genetic material into a diseased body in order to cure illness or alleviate symptoms as it takes its course. To date, a whole catalog of different strategies have been developed. Conceptually the simplest and most frequently used strategy is so-called gene addition for healing monogenetic diseases. It involves introducing an additional, so-called transgene to take over the function of the defective gene. The transgene DNA is introduced into the body with the help of vectors. At this time, only differentiated (somatic) cells qualify (somatic gene therapy) for the procedure. The genetic manipulation of germ line cells that would be passed on to the next generation is not justifiable on ethical ground and is prohibited in most countries. Gene therapy discoveries are misused in gene doping to enhance performance.

Gene transfer
In general, gene transfer refers to the exchange of genetic material between different organisms. In reproduction, it involves passing genetic material from one generation to the next of the same species (vertical gene transfer). Genetic material can also be transferred between organism of different species (horizontal/lateral gene transfer). In genetic engineering, gene transfer means the introduction of genetic material into the target cell for which a variety of techniques exist.

Genome
The sum total of an organism's genes. The entire human genome officially was completely decoded in 2003. As yet, not all gene functions are understood. Combining knowledge of how to decode genes with progress in identifying their functions leads to ever new possibilities and points of departure for gene modification.

Human image
Images of what it means to be human address questions of the nature of humankind, but in a tension between "what is" (what is a human being?) and "what ought to be" (what should the human man and woman be?). Expressed in human images is the interplay of basic worldviews, of the most varied norms and values that,

in turn, act in an orienting and action-guiding manner. The prevailing images of the human also influence the way enhancement techniques are viewed: when it comes to engineering the human being, the question of whether people should be conceived of as machines or rather as being made in a divine image can draw quite different answers. Inside sport, the human images become especially important, particularly with regard to the possibilities but also the limitations of physical performance ability. With the use of genetic engineering in sports, the anthropological attention here once more focuses on the proportion between naturalness and artificiality of these performances. Depending on the image of the human being, either the creative potentials made explicit by genetically engineered performance enhancement or the compromising of being human will be emphasized. Modern competitive sport hence opens a field of research, in which the conjuring up, the competition between, and the influence of traditional and contemporary views of the human image converge.

Judgment

Personal decision making and capacity to act, as well as discerning participation in societal debates, presume judgmental ability. It calls for weighing acquired knowledge, subjective and competing value sets in a reflective and grounded manner against each other in order to arrive in this way at a judgment about a situation, a conflict or a problem in a differentiated and responsible manner – and to do this even in situations under pressures brought on by potential social, media, or power relationships. As a personal skill assembled from subordinate capabilities, judgment can only be developed and fostered individually, albeit always within a social reference framework. Developing moral judgment and agency is regarded as a material precondition for developing a thoughtful understanding of technology/ Judgment, conceived of as a generalizable skill, can be transferred to any other area from the starting point of specific topic areas. Even if a highly-developed judgmental ability in no way automatically leads to a more mature moral action, it is nevertheless an essential precondition: the more highly developed judgment is, the more likely is it that moral judgment and moral action will be linked. The capacity for judgment can be developed especially through discursive examination of a topic.

Level playing field

This term defines a situation in which all participants in a competition formally operate under the same set of conditions to compete and win. The rules of a sport lay the groundwork for the level playing field, making it possible to compare the performances of those taking part in the actual event. The prohibition of doping is part of ensuring a level playing field. Should an athlete enhance his or her performance through genetic engineering in order to improve the chances of winning, the level playing field would no longer prevail.

Mass media
Mass media allow observing society from within society. So, modern dissemination media like the press, radio, TV, and the Internet can provide visibility in the public sphere to scientific topics, for example, make them accessible to the public, and supply topic-relevant background knowledge on them. Mass media treatment of the potentials and limits of genetic engineering enhancement practices in competitive sport currently ranges from factual reportages to dramatized hype.

Method (of gene transfer)
There are various methods for introducing genetic material into the body or the body's cells. The basic distinction is one of viral and non-viral gene transfers. Viruses are highly efficient at introducing genetic material into cells or, more precisely, into cell nuclei. Difficulties that arise from using viral vectors consist of immune reactions set off by the viral protein shell. This can lead to the vectors being "attacked" before the transgene reaches the target cell. Furthermore, transduced cells can be destroyed in an adaptive immune reaction so that the gene transfer effect is not sustained. Non-viral gene transfer, in which so-called plasmids are introduced into the cell nucleus, so far has proved to be very inefficient. A range of physical and chemical methods were developed to boost its efficacy. These consist, among others, of using slight electric fields (electroporation), high pressure (hydrodynamic transfection), or binding plasmids on cationic lipids or polymers. Even though these latter methods produce fewer side effects and dangers, they are not yet indicated for somatic gene therapy.

Morality
Moral questions relate to behavior, rules, principles, and virtues, etc. (see also "ethos of sport") of individuals, groups, or cultures (see Latin mos/ mores: moral/ mores). Moral appraisals fall into the area of normative ethics. At issue here are standards or goals oriented to what is morally good or right. In this connection, a rough distinction is drawn between deontological and utilitarian approaches that orient themselves by general validity conditions ("principles", "values") or action sequences ("the greatest good of the greatest number"). In competitive sport, the principles of level playing field and winning compete with each other, meaning that the individual athlete is challenged by his or her drive to win while simultaneously conceding the same right to competitors. The use of gene doping leads to a deception or cheating of the athletic opponent, meaning that the level playing field principle is violated. Lifting the restrictions on gene doping would

indeed restore the level playing field; however, with the consequence that the personal drive to excel would henceforth be tied to doping practices.

Myostatin

Most myostatin protein (MSTN) is produced in muscles and transported in the blood. This hormone acts as a negative regulator of skeletal muscle growth to control both the amount of muscle cells produced and the size of existing muscle fibers. Functional damage to this protein, e.g., through mutations in the MSTN gene, cause substantial muscle growth through production of a new muscle fibers (hyperplasia) and enlargement of existing muscle fibers. For this reason, methods for inactivating MSTN ("MSTN knockout"), which have already achieved success in animal gene therapeutic experiments, have become highly relevant, especially in sport types that emphasize strength.

NADA-Code

NADA stands for National Anti-Doping Agency and the NADA Code anti-doping regulatory framework is the most important set of anti-doping rules covering all sports on the national level. Among other contents, the NADA Code defines the concept of doping, stipulates which types of activities represent infractions of the anti-doping provisions, outlines the instructions for control and analytical procedures, and stipulates sanctions for violations. Since 2003, gene doping is also listed in the NADA Code as a prohibited method.

Naturalness

Naturalness includes a multitude of states and things that we postulate as givens and that came about without the aid of technology or by human intervention. Naturalness is something that humans have not influenced. However, defining naturalness in this way means that there is nothing left in daily life or in competitive sports that is natural. Strength training, ski jumping, or mountain biking are activities invented by us humans and that do not come naturally. Even running, strictly speaking, is not natural if we use cushioned running shoes or spikes. If we still insist on talking about the naturalness of sport, nevertheless, what we mean is that we take the human body and its limitations seriously and dispense with elaborate technical or medical aids. Manipulations in the form of genetically engineered enhancements along with other invasive influencing are not natural. Determining where naturalness ends and artificiality begins exactly is always also a societal negotiation process.

Prevention

In the battle against doping misuse in sport, two key strategies are the intensification and expansion of controls and sanctions or relying on educational-informative, preventive measures to appeal to the good sense of participating athletes. Prevention initiatives and programs like "Sport without Doping" (German Sports Youth/ Heidelberg Center for Doping Prevention) or "High Five – Together against Doping" (NADA) aim to achieve positive-preventive effects against doping misuse by making informative materials and brochures available. More recently, the prospects for success of such preventive measures were differentiated and commented on by (sports) science.

Public
The public (from Latin: public) principally means spectators and listeners of a great variety of events. It can be characterized in particular by their accessibility for anyone, voluntary attendance, and direct or indirect interactions without a duty to attend or interact directly or indirectly with the event. On-site spectators and media audiences can be differentiated in this regard. The latter can take an indirect interest in certain types of staging, for example, in sports. On-site spectators, in contrast, can directly exert an influence on the events, for example, with cheers, applause, and jeering. These individual behaviors can also always be fitted into moral categories (see also "morality") and range from positive to fairly negative examples of the public's conduct. Given the cited characteristics, competitive athletes frequently find themselves in a dependency relationship with the media public and the on-site public. Hence exists the danger that athletes will use gene doping in a performance enhancement to more fully live up to the public's expectations under the "higher, faster, farther" logic.

Risk
Every risk has two aspects: on the one hand, taking a risk heightens the chances and prospects for success in achieving a certain objective; on the other, the dangers involved also increase in proportion. When using performance-enhancing doping substances in sport, it means that, along with increasing one's own chances for success, the risk of being caught cheating (see also "Fraud") and of incurring health damage increases, with the consequence for the sport system of increased doping misuse in general (see also "Deviancy"). Although risk momentum is already inherent in the competitive idea of sport ("higher, faster, stronger"), with doping the dangers and drawbacks predominate for the participating actors as well as the sports system.

Self-doping
In self-doping, an athlete uses a prohibited substance or carries out a prohibited method on him- or herself. The offending athlete swallows the performance-

enhancing pills or injects the relevant substance – in brief, the athlete does the doping.

Signal transduction
Processes of signal transmission in cells are termed signal transduction. Stimuli, for example, hormones, trigger signals that are then transmitted from the cell surface to the cell interior. Signal transduction processes control gene transcription.

Sports ethos
Sports ethos, also termed "the spirit of sport," according to the WADA code describes a bundle of norms and values, such as honesty, fairness, justness, level playing field, or abiding by rules. In the ongoing ethics debate on forms of genetic engineering enhancement in competitive sports, the critics invoke the dissolution of these very values and norms.

Sport tribunal
Sport jurisdiction, which presupposes the existence of so-called sport tribunals, basically developed because national and international sports federations seek to keep disputes involving sports activities out of state courts. Reasons for this include that proceedings before sport tribunals cost less than those before state courts and that the proceedings tend take less time due to this type of court not being routinely overburdened like the state type. An added advantage is that a dispute in the sports field can be adjudicated by (sports) professionals. Consequently, a gene doping case is likely to be tried before a sport tribunal.

Technology assessment
Technology assessment reacts to the increasing technologization of modern society and the concomitant dependence on technology. Based on currently available knowledge, it prognosticates and weighs intended and unintended (retroactive and side) effects of technological innovation. In so doing, technology assessment runs up against the problem of an always uncertain future. In the context of potential genetically engineered enhancements in competitive sport, it brings into collective focus such problems as health risks for athletes, the evolving fundamental values in athletic competitions (see also "sports ethos"), as well as the ethical, legal, and anthropological deep structures of modern society.

Technology
Technology plays an ever increasingly important role in the differentiation of modern societies. Technology often is credited with the ability of reducing complex processes to simple and predictable cause-effect relationships. The hallmarks

of technology are above all efficiency, consistency, steering, and control. In medicine, for example, neurotechnology is making an impact on steering mechanisms of cognitive and emotional capabilities. Genetic engineering enhancement in sport consequently can be understood as a continuation of modern technological innovations. Originally developed for therapeutic purposes (gene therapy), gene doping holds out for competitive sport the prospect of having a kind of "push button technology" that promises to shift the natural limits on physical or mental potential. Like all new technology, enhancement through genetic engineering raises questions as to potential risks and side effects (see also "technology assessment").

Third-party doping
This type of doping is defined as the act by which a third party (doctor, coach, parents etc.) administers the prohibited substance to the athlete or performs the prohibited method on the athlete. The point is that the one doing the doping is doing it to someone else.

Transcription
Transcription is an important sub-process in gene expression describing the process by which DNA is copied into RNA.

Translation
Translation is a key sub-process in gene expression that describes what happens in protein biosynthesis, the process which turns information carried by mRNA into amino acid chains.

Verifiability (legal)
Verifiability that is legally sound (likely to stand up in court) refers to tallying up the characteristics of a situation or the correctness of an assertion or conjecture. Proving gene doping so that it will stand up in court is extremely difficult. Third-party introduced gene segments in blood or urine samples are not always detectable with direct detection methods. Verifiable detection would require a genetic examination of muscle cells preceded by extracting a muscle tissue sample. This, however, would be a profound infringement of the athletes' physical integrity (Article 2, Section 2 of the Constitution). Indirect detection methods (gene expression profile, blood passport) furnish alternatives. However, these methods can only determine that manipulation has occurred but not how it was done (either in substance or method).

Table of figures

	Captions and sources	Page(s)
1.	"Parties to the doping scandal"; Source: the authors	45
2.	"Genetically modified mouse I" Source: Lee (2007); PloS One. 29:2(8):e789	54
3.	"Genetically modified mouse II" Source: Lee (2007); PloS One. 29:2(8):e789	54
4.	"Options for gene transfers with somatic gene therapy" Source: Beiter & Velders (2012) DZS. Jg. 63, Nr. 5. p. 123	55, 62
5.	"Exon-Intron structure of genomic DNA (gDNA) and transgenic DNA (tDNA)" Source: Beitner et al. 2008, Exerc Immunol Rev. 2008; 14:73–85	60, 92
6.	"Cartoon 1" Source: Golombek; 2013	63
7.	"Cartoon 2" Source: Golombek; 2013	64
8.	"Conventional doping and gene doping compared" Source: the authors	65
9.	"Gene doping definition" http://www.nada-bonn.de/service-infos/; Source: the authors	66

All rights to the illustrations are reserved by the authors

www.ingramcontent.com/pod-product-compliance
Ingram Content Group UK Ltd.
Pitfield, Milton Keynes, MK11 3LW, UK
UKHW041438190426
11946UKWH00021B/15